RESTAURANT FINANCIAL MANAGEMENT

RESTAURANT FINANCIAL MANAGEMENT

A PRACTICAL APPROACH

Hyung-il Jung, PhD

Rosen College of Hospitality Management,
University of Central Florida, Orlando, Florida, United States

APPLE ACADEMIC PRESS

Apple Academic Press Inc.
3333 Mistwell Crescent
Oakville, ON L6L 0A2
Canada

Apple Academic Press Inc.
9 Spinnaker Way
Waretown, NJ 08758
USA

Exclusive worldwide distribution by CRC Press, a member of Taylor & Francis Group

International Standard Book Number-13: 978-1-77188-645-1 (Hardcover)
International Standard Book Number-13: 978-1-315-14739-0 (eBook)

Library and Archives Canada Cataloguing in Publication

Jung, H (Hyung-il), author
Restaurant financial management : a practical approach / H. Jung, PhD (Rosen College of Hospitality Management, University of Central Florida, Orlando, Florida, United States).

Includes bibliographical references and index.
Issued in print and electronic formats.
ISBN 978-1-77188-645-1 (hardcover).--ISBN 978-1-315-14739-0 (PDF)

1. Hospitality industry--Finance. 2. Restaurant management. I. Title.

TX911.3.F5J86 2018 647.95068 C2018-903678-8 C2018-903679-6

CIP data on file with US Library of Congress

Apple Academic Press also publishes its books in a variety of electronic formats. Some content that appears in print may not be available in electronic format. For information about Apple Academic Press products, visit our website at **www.appleacademicpress.com** and the CRC Press website at **www.crcpress.com**

CONTENTS

ABOUT THE AUTHOR

Hyung-il Jung, PhD
Associate Lecturer, Rosen College of Hospitality Management,
University of Central Florida, Orlando, Florida, USA,
Tel.: (407) 903-8175, E-mail: hiljung@ucf.edu

Hyung-il Jung, PhD, has been teaching at the Rosen College of Hospitality Management of the University of Central Florida since 2005. Before joining the Rosen College, he taught at Roosevelt University in Chicago. Throughout his academic career, he has been emphasizing for students the qualitative interpretation of quantitative data by teaching hospitality financial and managerial accounting, hospitality finance, and feasibility studies.

His interest in this line was developed during his days at Florida International University when he was studying for his master's degree, after which he spent almost ten years in the industry to practice and cultivate his ideas further, working as a systems designer, an operations analyst, and the controller of a few nationwide foodservice companies that served convention centers and big sporting events, such as ball games, Winston Series Stock Car Races, and National Final Rodeo games. His industry career helped him advance his specialty of liaising between operators and the back of the house with structured assistance that combined accounting information and operational data.

Dr. Jung earned his PhD degree from Virginia Tech. Back in academia, Dr. Jung has been working to accomplish his goal of laterally connecting the cores of Financial Accounting, Managerial Accounting, and Finance into one complete set of applicable business models to help students of Hospitality Management learn the subjects from practical perspectives for their career.

LIST OF ABBREVIATIONS

ACP	average check price
AICPA	American Institute of Certified Public Accountants
CAPM	capital asset pricing model
COS	cost of sales
EBIT	earnings before interest and taxes
EBT	earnings before Taxes
FC	fixed cost
FV	future value
GAAP	generally accepted account principles
NPV	net present value
POS	point-of-sales
PV	present value
ROA	return on assets
ROE	return on equity
ROI	return on its investment
ROS	return on sales
TVM	time value of money
VC	variable cost
WACC	weighted average cost of capital

PREFACE

Numerous books introduce the techniques of accounting and finance in the realm of restaurant management. All of them have their own unique strengths. Few of them, however, seem to provide practical guidance to readers regarding how to apply the concepts of accounting and finance to real-life business activities. Most of the existing books emphasize rather the technical aspect from the operational perspective. This limited approach only helps readers develop a partial picture of a business that misses the impacts of financing activities. Financing activities of a business requires investment decisions of financial suppliers, which impose the ultimate risk to the business itself.

This book deviates from many others by providing readers with a framework that consists of three steps of applying techniques of accounting and finance to evaluate a restaurant business. The first part introduces how to consolidate major activities of a restaurant business into accounting information. Then, in the second part, it explains how accounting information is analyzed and used to project the future. Finally, the third part introduces the methods of projecting the future and determining the current value of a restaurant business. Using this approach, it is hoped that readers can develop useful knowledge regarding how to relate accounting and finance to real-life businesses.

This book uses an imaginary restaurant business as an example to demonstrate a series of relevant business activities to guide readers to understand how those activities are transformed into meaningful information. Most of the operating data used in this book have been slightly modified from the actual ones provided by one restaurateur, who does not want to be exposed. By following this process, the author desires to help readers develop a bird's eye view over a restaurant business.

As mentioned above, this book is divided into three large parts. The first part is about financial accounting techniques that generate necessary information into a set of financial statements. In this section, only several types of representative business transactions are introduced in simplified formats. The second part introduces a few necessary sets of analytic tools (or ratios) to measure the effectiveness and the efficiency of the business activities. This section also introduces how those tools are used in forecasting and planning. In the last part, the concept of cash flows is introduced with the techniques of valuing a business. The reason for introducing the concept of cash flows in the last part is that the true value of a business is always determined by its capability of generating cash for its owners.

In this context, this book introduces the ultimate use of accounting information in a financial framework that measures the value of a restaurant business. It is the wish of the author that readers of this book will develop comprehensive knowledge regarding how to use accounting information to achieve the financial goal in a restaurant business.

FOREWORD

Those of us in the academic world of hospitality management (whatever the title of our college or department) are preparing our students to take up operational roles in the burgeoning industry. This takes a practical approach to learning. Finally we have an accounting/finance book that takes that practical approach.

We hear the hue and cry for more STEM education, but we can't forget that while STEM is important, business is of paramount importance and knowing the financial systems of business is an imperative. What Dr. Jung has done is give us a textbook that emphasizes the practical skills that operators need to have to be leaders in the industry.

I wish that I had this book to learn from when I was a young restaurant manager. It would have made my journey as an operator much easier. It wasn't until I learned the language of the finance world that I really became successful. Having been an executive with three major corporations (almost twenty years with Walt Disney World Co.), I can tell you that the lessons of this text are crucial for every leader's ability to achieve their goals.

Dr. Jung has laid out the book in an easy-to-follow manner that allows the student to understand the building blocks of accounting and finance. The case study approach allows the learner to absorb the information and realize how to apply it in the world that they will soon be entering. Whether it be revenue, expenses, ROI, or budgeting, the lessons of this text will soon become everyday living for our students. They need this information.

Duncan R. Dickson, EdD
Associate Professor, Tourism, Events, and Attractions,
Rosen College of Hospitality Management,
University of Central Florida, Orlando

PART I

HOW TO ORGANIZE COMMERCIAL ACTIVITIES INTO FINANCIAL INFORMATION

CHAPTER 1

INTRODUCTION: THE ROLE OF ACCOUNTING IN A BUSINESS

1.1 A BUSINESS AS A SEPARATE ENTITY

It is important to realize that a business is an independent entity that conducts continuously many different activities of its own to achieve the final goal of adding value to the original investments made by the owners. In this context, a business must be created in the beginning with initial investments made by a person or a group of people who want to make their original investments increase in the long run.

Once the initial investment is made, a business firm is created, and the investors become the owners of the business. From this point, however, the business (or firm) is considered as an independent entity that conducts its own activities. As an independent entity, the activities of the business must be recorded separately. This concept is defined in accounting as the principle of "Business Entity" or "Economic Entity." The following equation presents the financial position of a firm. "Assets" represent the financial value of the firm, while the "Liabilities" and "Equity" refer to the amount contributed by two different groups of financial suppliers. More details are explained in the following sections.

Assets = Liabilities + Equity (A = L + E);
the Fundamental Accounting Formula

1.2 ASSETS AND EQUITY

Now that the business is considered as an independent entity from the owner(s), the initial investment made by the owners becomes a business transaction, and it must be recognized. It is called the "Equity" of the firm. It is also referred to as "Owners' Equity." If the business is incorporated, it is called "Stockholders' Equity" or "Shareholders' Equity" because the owners receive common stock as their ownership certificates in exchange for their financial investments. The invested amount, which was given to the business, becomes the property of the firm for its own use. In other words, the entire amount received becomes the resource that belongs to the firm itself; and it is called the "Assets" of the business. At this point, the company's Assets, which are usually in the form of cash, are the same as its Equity in amount (Assets = Equity). More specifically, the firm has cash (which is its Assets), and the business recognizes it as Equity that was invested by the owners. It is important to notice that this activity of investing has been recorded in two different forms– one as the firm's Asset and the other as Owners' Equity. Both have increased.

Owners have put their own (personal) resources into a firm with the expectation that their investments will grow in time. For the owners, this is an investment activity. For the firm, however, it is a financing activity of obtaining necessary capital to start necessary work. When a business obtains its necessary funds from owners, it is called "equity financing."

With the initial funds (or capital) obtained, the firm starts spending it for necessary material such as buildings (or space to work), equipment, furniture, and other resources to prepare for its operations. It also has to prepare its workspace to accommodate customers and make it look attractive and convenient for its customers and employees. All these activities result in adding additional assets, and they require a lot of spending. Spending for these is considered "internal investing activities of the business." These activities are often introduced as "resource allocation," which results in adding other forms of assets. In most cases, these "internal investing activities" or "resource allocation" require additional funds that must be borrowed or paid later. In either situation, the business ends up owing to other stakeholder group, which is explained in the following section.

1.3 ASSETS AND LIABILITIES

As pointed out, during the process of a company's internal investing activities (or resource allocation), the business may need to take loans for more funds or it may end up owing unpaid fees to many vendors and service providers. These debts are called the "Liabilities" of the business. In these situations, loans and other debts increase the firm's liabilities. It is important to notice that the increase of liabilities was caused by the firm's activities of increasing its Assets. The equation just for these is now $A = L$, indicating assets increase with the increase of the firm's liabilities.

$$\boldsymbol{A = L + E}$$

When the two different equations introduced above ($A = E$ and $A = L$) are combined together, the equation becomes the fundamental accounting equation of $\boldsymbol{A = L + E}$ (Assets = Liabilities + Equity) that is the backbone of all business status and activities. Up to this point, a firm has been conceptually created and it has increased the volume of its assets. However, the increase of its assets has been achieved only through the financing activities of accepting owners' equity investments and of borrowing. Nothing has been achieved by its operating activities of serving customers.

1.3.1 Examples of the Description Above (Owners' Investments)

Let's create a hypothetical example of opening a small business and apply accounting concepts introduced above as follows:

> **Example 1.1:** *Rachel and Brad, a happily married couple, wanted to have their own restaurant business for long. Finally, they are fully prepared and are opening their own restaurant business. They incorporated their business as "R&B Grill." Their initial investment was $450,000 in cash. In return, they received common stock. Both of them will manage the business full-time. From now on, "R&B Grill" will be used for the name of the business.*

The Example 1.1 above is about opening a business. According to the Business Entity principle, accounting only records the monetary value exchanges of the business. According to the description, Rachel and Brad opened their own business and became the owners (or investors). At the same time, they also have become employees (managers) of their business (R&B Grill). R&B Grill as a business entity has received $450,000 of cash from the owners (=investors). This cash has become the company's property. The property owned by a business becomes an asset item.

Now that the business (the R&B Grill) has received $450,000 of cash, the company's Assets have increased by financing through issuing common stocks. In other words, the investors have invested their money into it and became its owners. Owners' invested money is called the "Equity" of the company. In conclusion, R&B has increased its Asset (in cash) by $15,000 and it now recognizes its Equity by $450,000. Thus, the company's Asset ($450,000) = Owners' Equity ($450,000).

This is a big picture of the business in its financial position. Assets, Liabilities, and Equity are group names that include many items that are called "accounts." Accounting explains business situation with the details of specific accounts. For example, the Assets include such accounts as "Cash," "Accounts Receivable," "Inventories," "Prepaid Expenses," "Furniture, Fixture, and Equipment (FF&E)," and many others. The common nature of all these accounts is that they are the resources owned by the business for its own use. The account group of Liabilities includes "Accounts Payable," "Accrued Expenses," "Notes Payable," and many others. As these titles indicate, the common nature of them is that they are all debts to pay. Finally, the account group of Equity includes "Common Stocks," "Preferred Stocks," "Treasury Stocks," and "Retained Earnings." Oftentimes, it may present a long title that says "Paid-in Capital in Excess of Par." The descriptions of these titles will be explained later in the book.

The example also says that the company has issued Common Stocks to the investors. As introduced above, "Common Stock" is an account that belongs to the equity account group. The actual accounting description of this transaction is that the company's cash in Assets has increased by $450,000, and its common stocks in Owners' Equity have increased (or been recognized) by $450,000. Let's continue with a loan using the same example. When a business borrows by taking loans, two impacts are made as follows: its Assets (in cash) increase and its liabilities (loan = debt) also increase.

Example 1.2: *Rachel and Brad (as the managers of R&B) decided to take a loan of $150,000 from a local bank in order to purchase necessary furniture and equipment. The bank has approved their loan application and issued a loan of $150,000 for 3 years at 12% annual interest.*

This transaction has brought in another batch of cash ($150,000) to the business, increasing the balance of cash, resulting in an increase in assets. Let's take a look at the source of this cash increase. It was caused by the loan, which is the nature of liabilities. In conclusion, this transaction has increased the company's assets and it also has increased its liabilities, which is represented in equation as Assets (increase) = Liabilities (increase).

This transaction mentions the interest on the loan. Interest is an expense that is incurred as time goes by. At this point, however, there is no need to record interest amount. Interest expenses will be incurred and calculated at the closing of one accounting period such as 1 month.

With no interest to be calculated, this transaction only involves an asset account and liability account. As assets and equity have been narrowed down in previous transactions to the account titles of "Cash" and "Common Stocks," liability in this example needs to be assigned as an appropriate account. The transaction says that this loan is to be paid in 3 years. When a liability item (debt) is allowed to be paid for the period longer than 1 year, it is considered as a long-term debt. The most commonly used account title for such a case is "Notes Payable." The equation in this example, therefore, can be represented as follows:

Assets (cash) increase $150,000 = Liabilities (in Notes Receivable) increase $150,000

Let's put both examples together to see how each equation explains the company's situation. When both examples are combined, the company's cash is now $600,000, which is the total of $450,000 of initial investment (equity) and $150,000 obtained through a loan (Liability). The final equation of Assets ($600,000) = Liabilities ($150,000) + Equity ($450,000) explains the amount of the company's total assets and the sources of the entire Assets. It is presented in the following display:

	Assets	=	Liabilities	+	Equity
Example 1.1:	Cash $450,000	=			Common Stock $450,000
Example 1.2:	Cash $150,000	=	Notes Payable $150,000		
Total:	Cash $600,000	=	Notes Payable $150,000	+	Common Stock $450,000

Now, R&B has $600,000 in its Assets, which is made of $150,000 of Liabilities and $450,000 in Equity. Another interpretation of this is that the company's financial structure is made of 25.0% of liabilities ($150,000 of debt/$600,000 of assets x 100 = 25%) and its equity is 75.0% of its assets ($450,000 equity/$600,000 assets x 100 = 75%). When accounting information is presented this way, it is called the financial position of the company. Let's practice another similar example as follows:

Example 1.3: *One of their friends, Jane, shows strong interest in their new business and decided to join as an investor. She decided to invest $200,000 and received Common Stock.*

This transaction is identical in the nature to the first example. The only difference is that the investor is another person. To the business, cash increased by $150,000, and it is necessary to recognize additional owners' equity of $200,000 under the title of "Common Stock." For record-keeping purpose, however, the common stock issued to two different groups may have different subtitles such as "Common Stock – Rachel and Brad" and "Common Stock – Jane" with specific amounts contributed by each. Detailed recording process will be explained later in this section. The impact of this transaction is as follows:

Assets	=	Liabilities	+	Equity
Cash $200,000	=			Common Stock $450,000

The company's financial position at this point is:

	Assets	=	Liabilities	+	Equity
Example 1.1:	Cash $450,000	=			Common Stock $450,000
Example 1.2:	Cash $150,000	=	Notes Payable $150,000		
Example 1.3:	Cash $200,000	=			Common Stock $200,000
Total:	Cash $800,000	=	Notes Payable $150,000	+	Common Stock $650,000

1.4 RECORDING PROCESS OF FINANCIAL ACCOUNTING FOR MORE DETAILS OF BUSINESS ACTIVITIES

The conceptual big picture of accounting has been introduced so far with examples of the increase of Assets, Liabilities, and Equity. In fact, the result of the three examples comes through more detail-oriented steps known as the recording process, which involves specific account titles that belong to the larger account groups of Assets, Liabilities, and Equity. As introduced in the beginning, the Asset account group includes account titles such as "Cash," "Accounts Receivable," "Inventories," "Prepaid Expenses," "Furniture, Fixture, and Equipment (FF&E)," and many others. The business owns them for its use to conduct necessary activities. The account group of Liabilities includes titles such as "Accounts Payable," "Accrued Expenses," "Notes Payable," and other similar accounts. The company must pay these debts at a later date to creditors. Finally, the Equity account group includes "Common Stocks," "Preferred Stocks," "Treasury Stocks," and "Retained Earnings" as introduced earlier.

1.5 JOURNALIZING

Journalizing is a process of allocating the accurate amounts and impacts (increase or decrease) of individual transactions into relevant accounts. Each transaction has at least two impacts. As shown earlier, the Example 1.1 has an impact in increase of Assets and another increase of Equity. The Example 1.2 shows an impact in increase of Assets and an increase of Liabilities. The Example 1.3 has an impact in increase of Assets and another increase of Equity. More specifically, the Examples 1.1 and 1.3 use the account title of Cash to record the increase of Assets and uses the account title of Common Stock to record the increase of Equity. Similarly, the Example 1.2 uses the title of Cash to record the increase of Assets as the result of the transaction, and the title of Notes Payable to record the increase of Liabilities.

In total, the three examples (1.1–1.3) introduced above have presented the increase of Cash, Common Stock, and Notes Payable under the equation Assets = Liabilities + Equity. It is suggested now to consider the equal sign ("=") in the equation as a dividing line. Then, the entire space is divided into the left side and the right side. In the examples, Cash increase was put on the left side under the Assets, while the increases of Common Stock and Notes Payable are presented on the right side under Equity and Liabilities, respectively. In this context, the simple equation of $A = L + E$ also represents an important concept of correct sides that indicate the increase of each account group. When Assets increase, the impact must be recorded on the left side and when Liabilities or Equity increase, those impacts must be recorded on the right side. More of this will be explained in the following section.

1.6 DEBIT AND CREDIT

In accounting, the left side is named "Debit" and the right side is named "Credit." Although they mean nothing but each side of left and right, when the concept of Assets, Liabilities, and Equity is applied to these, they start to carry a specific impact of either "Increase" or "Decrease" of each account group. As introduced in the examples, the increase of Assets is presented on the debit (left) side. On the other (credit or right) side, the increase of Liabilities or Equity is presented. Naturally, the decreases simply take the opposite side. The decrease of Assets is presented on the right (credit) side, while the decrease of Liabilities or Equity is presented on the left (Debit) side.

This way, the equation of Assets = Liabilities + Equity can also be used to indicate the increase or decrease of each account group. It must be remembered that the total amount presented on the debit side must be the same as that of the credit side. This rule must be applied to individual transactions and the total as well. The following is the summary of this concept with the conventional format of recording journal entry, followed by a diagram that illustrates the dynamics of business transactions on the Assets, Liabilities, and Equity.

Summary of Debit and Credit

Debit entry		Credit entry	
Impact	**Format**	**Impact**	**Format**
Assets Increase	(acct. title $$,$$$)	Assets Decrease	(acct. title $$,$$$)
Liabilities Decrease	(acct. title $$,$$$)	Liabilities Increase	(acct. title $$,$$$)
Equity Decrease	(acct. title $$,$$$)	Equity Increase	(acct. title $$,$$$)

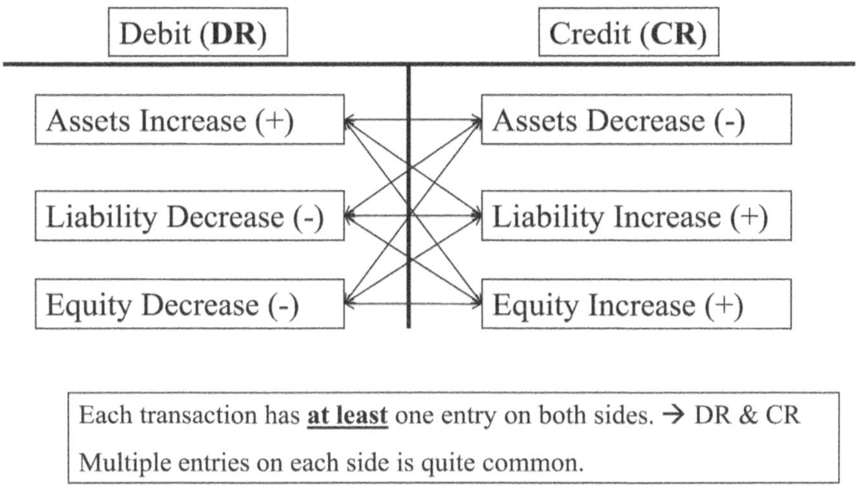

Figure 1.1 Double Entry System: Rules of Journal Entry Debit and Credit with Assets, Liabilities, and Equity.

The summary and Figure 1.1 presented above show the location of each record depending on its impact on the relevant account group. When one record is put on one side, there must be at least another record put on the other side. Let 's reprocess the three examples using the new technique.

In the Example 1.1, the initial investment of $450,000 made by Rachel and Brad was explained to increase the company's assets under Cash account, and it was recognized as Owners' Equity using the account title of Common Stock. In the journalizing process, it will be recorded as follows:

Table 1-1(A) Example 1-01 Journal Entry

Debit			Credit		
Impact	Account	Amount	Account	Amount	Impact
A+	Cash	450,000	Common Stock – Rachel & Brad	450,000	E+

Description: Rachel and Brad made an investment.

The impact of this transaction is the increase of assets on the debit side and the increase of the equity on the credit side. Most accounting books do not mention impacts. However, it is very important, particularly for managers and operators, to understand accounting this way to develop comprehensive knowledge on how to view business situations through accounting records. The second transaction of obtaining a loan of $150,000 will be journalized as follows:

Table 1-2(A) Example 1-02 Journal Entry

Debit			Credit		
Impact	Account	Amount	Account	Amount	Impact
A+	Cash	150,000	Notes Payable	450,000	L+

Description: A bank loan has been made.

The next example of the additional investment from Jane of $200,000 (1-3) will be journalized as follows:

Table 1-3(A) Example 1-03 Journal Entry

Debit			Credit		
Impact	Account	Amount	Account	Amount	Impact
A+	Cash	200,000	Common-stock-Jane	200,000	E+

Description: Additional investment has been by Jane.

Now, all three initial examples are completely journalized in a simplified format. In practice, journal entries must have other supporting information, such as date and record number, to name a couple. It must also be noticed that each entry has brief explanation of the transaction. "Journalizing" is just like keeping a diary of the activities of a business. Most businesses use account numbers these days instead of descriptive account titles, thanks to the computerized system that stores and matches numbers to relevant titles.

1.7 POSTING TO INDIVIDUAL *T*-ACCOUNTS

Once a journal entry is complete, the record must be transferred to individual accounts (called *T*-accounts) separately. This process is the step of consolidating widely scattered data into individual accounts. The volume of journal entries, in any business, must be too large and disorganized to pinpoint the financial situation at any time. To provide timely and accurate information such as the cash available or the amount to be paid to a certain vendor, the manager of the business must be able to have specific information quickly. This task is conducted at the second step of the recording process – posting. The three examples introduced so far will be continuously used to explain the posting process.

The journal entry of the first example is reproduced below:

Example 1-01 Journal Entry

Debit			Credit		
Impact	**Account**	**Amount**	**Account**	**Amount**	**Impact**
A+	Cash	450,000	Common Stock – Rachel & Brad	450,000	E+

Description: Rachel and Brad made an investment.

In the posting process, each *T*-account title presented in the journal entry must be located. Traditionally, each account is used to take the form of "T" because of the need to record debit and credit entries. Although today's computer use has eliminated the physical use of *T*-accounts, the concept of debit and credit is still used to indicate the increase or decrease. Each and every business has its own list of account titles.

Once the *T*-account is located, the journal entry is simply "posted" to the *T*-account. The debit entry of "Cash 450,000" in the Example 1.1 is posted to the same "debit" side of the Cash *T*-account. In this process, by the way, only the amount is posted since the *T*-account is that of "Cash." In the same manner, the credit entry of "Common stock – Rachel & Brad 450,000" is posted to the *T*-account of the "Common stock – Rachel

& Brad" onto the credit side. Once this is done, they will look like the following:

Table 1-1(B) Example 1-01 Posting

Cash		Common Stock – Rachel & Brad	
Ex 1-01 450,000			450,000 Ex 1-01

As presented in Table 1.1(B), the Cash *T*-account now shows a debit record of $450,000 while the Common stock – Rachel and Brad *T*-account shows a credit record of $450,000. The "Example 1.1" in each *T*-account is the reference number to identify the relevant journal entry when verification is needed. In practical accounting work, all necessary reference numbers are usually presented as identifiers due to the complexity created by the volume of data. In this book, however, a simplified format is used for the convenience sake. Just like the individual journal records, the total debits of all *T*-accounts must equal the credits. Also, it must be noticed that all the records of one journal entry on both sides are posted to the same side of relevant *T*-accounts. In other words, individual journal entries are recorded onto each side (debit or credit) of relevant *T*-accounts and the total amount of debits equals that of the credits. The following is the continuation with the other two examples. It must be noticed that the same Cash *T*-account will be continuously used for following entries of Cash. However, whenever a new account title is used, new *T*-account must be introduced.

The second example is the loan of $150,000 from a Bank ABC, which is journalized as follows:

Example 1-02 Journal Entry

Debit			Credit		
Impact	**Account**	**Amount**	**Account**	**Amount**	**Impact**
A+	Cash	150,000	Notes Payable	450,000	L+

Description: A bank loan has been made.

The posting mechanism remains the same. The debit entry goes to the left side of the relevant *T*-account and the credit entry to the right.

Table 1-2(B) Example 1-02 Posting

Cash			Notes Payable – Bank ABC		
Ex 1-01	450,000			150,000	Ex 1-02
Ex 1-02	150,000				

The Cash *T*-account now shows the Example 1.2-a record on the debit side, while a new *T*-account of "Notes Payable – Bank ABC" is showing a credit record of $150,000. By looking at the Cash *T*-account, we can easily find that Cash has increased twice by viewing the debit side and the total amount of Cash on hand at this point is $600,000. Let's complete this process with the last example, which is journalized as follows:

Example 1-03 Journal Entry

Debit Impact	Account	Amount	Credit Account	Amount	Impact
A+	Cash	200,000	Common-stock-Jane	200,000	E+

Description: Additional investment has been by Jane.

This journal record keeps using the Cash *T*-account, but the Common Stock –Jane's *T*-account must be created. The following is the result of posting:

Table 1-3(B) Example 1-03 Posting

Cash			Common Stock – Jane		
Ex 1-01	450,000			200,000	Ex 1-03
Ex 1-02	150,000				
Ex 1-03	200,000				

There are four *T*-accounts. Of those, Cash belongs to the assets account group, Notes Payable belongs to the liabilities account group, and finally the two Common Stock accounts belong to the equity account group. As this is only the beginning stage, the business only has increasing records of each account group. The total Cash amount is $800,000, while the amount of Notes Payable is $150,000, and finally, the two Common Stock accounts combined together show $650,000 in equity. The totals of individual *T*-accounts also show Assets (total) = Liabilities (total) + Equity (total). This is the beginning financial position of R&B Grill. Rachel and Brad, as managers, can start building up their business with necessary furniture, equipment, and other items using the cash to prepare for the real business of serving customers.

Table 1-04 *T*-Accounts with Initial Balances

Cash		
Ex 1-01	450,000	
Ex 1-02	150,000	
Ex 1-03	200,000	
	800,000	

Notes payable		
	150,000	Ex 1-02

Common stock – Rachel & Brad		
	450,000	Ex 1-01

Common stock – Jane		
	200,000	Ex 1-03

1.8 PREPARING THE INITIAL BALANCE SHEET OF R&B GRILL

Before moving into the next phase of purchasing furniture, equipment, and other material, it is helpful to consolidate all the information into a summarized statement that shows the information in a more structured manner. First, we need to calculate the balance of each *T*-account and post them into a financial statement that summarizes the situation. The following is the same *T*-accounts with the total amount shown on the bottom with a double underline.

Table 1-05 The Balance Sheet on the First Day

The Balance Sheet R&B Grill, Inc. On January 01, 2XX1			
Assets:		**Liabilities & Equity**	
		Liabilities	
Cash	800,000	Notes payable	150,000
		Equity	
		Common stock	
		– Rachel & Brad	450,000
		Common stock	
		– Jane	200,000
Total Assets	800,000	Total Liabilities & Equity	800,000

Almost all the textbooks on financial accounting introduce the step of preparing a Trial Balance, which is an error-checking procedure between the *T*-account balance calculation and the preparation of financial statements. In this book at this point, however, the Trial Balance step is eliminated assuming all calculations are correct. After all, there are only four simple *T*-accounts. The financial statement being prepared is the Balance Sheet, which shows the company's financial position at this time. It is prepared by posting the balance of each *T*-account into its relevant account group. That is, Cash *T*-account balance is transferred to the assets, Notes Payable balance is transferred to the liabilities and finally, the balances of the Common Stock *T*-accounts are transferred to the equity. The final view of the statement looks like the following Table 1-05 "the Balance Sheet on the first day."

Table 1-05 The Balance Sheet on the First Day

The Balance Sheet R&B Grill, Inc. On January 01, 2XX1			
Assets:		**Liabilities & Equity**	
		Liabilities	
Cash	800,000	Notes payable	150,000
		Equity	
		Common stock	
		– Rachel & Brad	450,000
		Common stock	
		– Jane	200,000
Total Assets	800,000	Total Liabilities & Equity	800,000

As shown on the bottom of Table 1.5, the company currently has $800,000 of assets in cash. The right side of the Balance Sheet tells that the total amount of assets has been obtained from two sources that are liabilities and equity. In other words, the company has financed $800,000 of asset (in cash) by mixing the loan and stock issuance.

1.9 ACTIVITIES OF RESOURCE ALLOCATION (INTERNAL INVESTING ACTIVITIES TO PREPARE FOR BUSINESS)

Now, the management of R&B Grill is ready to start working on preparing to open the business. It starts with securing a place to conduct its business. Let's assume that the company leases an empty building as described in the next example (Example 1.4).

> **Example 1.4**: *There was an empty building available for lease that used to be a restaurant in town. On Jan. 1, a lease contract was signed for the building. Rent is set at $15,000 per month. The landlord demanded a $100,000 deposit for future rent in*

advance. Rachel and Brad agreed and leased the place for 5 years. The payment was recorded in the account named "Prepaid Rent."

In this transaction, R&B Grill has paid cash to secure a place for its operations. This spending is not an expense because the space has not been used yet. In other words, the Rent Expense has not been incurred yet. The payment was to create a deposit account that purchases a legal right to use the space.

The amount of cash paid has decreased the amount of its cash on hand (asset decrease) and the decrease of cash must be journalized in the credit side. The debit side records, though, the deposit for the "legal right" to use the place. This deposit account title is "Prepaid Rent." In the Balance Sheet, this kind of asset is consolidated into the group of "Prepaid Expenses." (Many public restaurant companies' balance sheets show the account title of "Prepaid Expenses.") The journal entry and impacts will look like the following:

Table 1-06(A) Example 1-04 Journal Entry

Debit			Credit		
Impact	Account	Amount	Account	Amount	Impact
A+	Prepaid Rent	100,000	Cash	100,000	A–

The journal entry shows the impact on each side, which is not a usual practice in accounting. "A+" stands for an increase in Assets, while "A-" indicates a decrease in Assets. This presentation shows only to guide the readers to develop necessary knowledge to use accounting information in the long run. The interpretation of the transaction is that the company has obtained an asset item by paying cash. The "Prepaid Rent" gives R&B Grill the right to use the space for its business. Remember that "Prepaid Rent" is not Rent Expense yet. As time goes by, the company will spend its deposited amount to pay for the monthly rent. Only then, the amount spent each month will be recorded as monthly Rent Expense. All other Prepaid Expenses work the same way.

When this journal entry is posted, the relevant *T*-accounts are Cash and Prepaid Rent. They will look like the following:

Table 1-06(B) Example 1-04 Posting (Cash & Prepaid Rent T-accounts)

Cash				Prepaid Rent		
Ex 1-01	450,000	100,000	Ex. 1-04	Ex. 1-04	100,000	
Ex 1-02	150,000					
Ex 1-03	200,000					

The Cash *T*-account shows one record ($100,000) on the credit side, indicating there was an outflow of cash of $100,000. With this record, the company's cash balance drops from $800,000 to $700,000. However, it has obtained another asset (Prepaid Rent) in the same transaction. Although its cash balance has decreased, its total asset value still remains the same. Simply, R&B Grill has invested its cash to secure the business site. If the Balance Sheet is prepared at this point, it will look like the following:

Table 1-07 The Balance Sheet After Lease Contract (Example 1-04)

The Balance Sheet R&B Grill, Inc. On January ##, 2XX1 – after Lease Contract			
Assets:		**Liabilities & Equity**	
		Liabilities	
Cash	700,000	Notes payable	150,000
Prepaid Rent	100,000		
		Equity	
		Common stock	
		– Rachel & Brad	450,000
		Common stock	
		– Jane	200,000
Total Assets	800,000	Total Liabilities & Equity	800,000

The next step covers the preparation of the business by installing equipment and furniture along with necessary contract works, such as plumbing, electricity, building cabinets and storage space, and other layout that supports operations. The next example (Example 1.5) introduces a simplified scenario of such transaction.

Example 1.5: *Rachel and Brad decided to purchase a new set of furniture and fixture for the dining room and new sets of kitchen equipment. Following the industry norm, these are consolidated into an asset account of FF&E (Furniture, Fixture, and Equipment). The Manufacturer's Suggested Retail Price (MSRP) was $550,000, but the entire set was on sale for $500,000. One half was paid in cash up front, and the other half was put on account for 2 years. The vendor decided not to charge any interest on the balance if R&B promises future purchases from them. Rachel and Brad agreed to do so. The FF&E will be used for 5 years.*

This type of spending is called a capital expenditure, which stands for an internal investment. This should not be mistaken for an expense. The definition of an expense will be provided later in this chapter. The huge spending in this transaction has brought in a large amount of equipment and furniture. They are all asset items. "FF&E" is the commonly used account title for these. The purchase of the FF&E in this transaction includes cash spending (Assets decrease; A-), and an unpaid amount that has become a debt (Liabilities increase; L+). Another fact to notice is the two different amounts for the FF&E. The MSRP is $550,000 and the purchase price is $500,000. This is an example of an Accounting Principle called the Cost Principle or the Historical Cost Principle. Under this principle, accounting only records the actual amount paid for any items. The MSRP or any other values are irrelevant in accounting. Thus, the amount of $500,000 should be recorded. Following the rules introduced earlier, the journal entry of this looks like the following:

Table 1-08(A) Example 1-05 Journal Entry

Debit			Credit		
Impact	Account	Amount	Account	Amount	Impact
A+	FF&E	500,000	Cash	250,000	A–
			Notes Payable	250,000	L+

From now on, the description of each journal entry will be eliminated. As mentioned, this transaction includes two entries to the credit side. A journal entry that has multiple records at least on one side is called a "composite entry" in technical term. The interpretation of this journal entry is that the company has obtained an asset item (FF&E) by paying cash for one half and incurring a liability for the other half. The relevant *T*-accounts, after posting, are presented in the following Table 1.8(B).

Table 1-08(B) Example 1-05 FF&E Purchase Posting

Cash				FF&E			
Ex 1-01	450,000	100,000	Ex 1-04	Ex 1-05	500,000		
Ex 1-02	150,000	250,000	Ex 1-05				
Ex 1-03	200,000						

Notes Payable			
		150,000	Ex 1-02
		250,000	Ex 1-03

It must be noticed that the Notes Payable *T*-account is now updated to include the current example (Example 1.5) in addition to the earlier record of $150,000. Now, R&B Grill has increased its liabilities to $400,000. This result will change the financial position of the company as follows:

Table 1-09 The Balance Sheet After FF&E Purchase (Ex. 1-05)

The Balance Sheet R&B Grill, Inc. On January ##, 2XX1 – after FF&E purchase			
Assets:		**Liabilities & Equity**	
		Liabilities	
Cash	450,000	Notes payable	400,000
Prepaid Rent	100,000		
		Equity	
FF&E	500,000	Common stock – Rachel & Brad	450,000
		Common stock – Jane	200,000
Total Assets	1,050,000	Total Liabilities & Equity	1,050,000

The new balance sheet in Table 1.9 shows three items of assets owned by R&B Grill, which is $1,050,000 in total. The total assets have been financed by $400,000 of Notes Payable (Liabilities) and by $650,000 of the investment made by the owners (Equity). As a result, this company owns assets of $1,050,000. Its creditors claim $400,000 on the assets, while the investors can claim the rest that is $650,000. In summary, the liabilities and equity of a business is also described as the claim of each stakeholder group – creditors and owners – on the entire assets of the business.

Even after installing equipment and furniture, R&B is not ready to open its door to accommodate customers. The company still has to go through a few more steps that include purchasing fire insurance. The next example (Example 1.6) introduces the transaction of fire insurance purchase.

> **Example 1.6:** *Rachel, as the operating manager, purchased fire insurance for 1 year. The annual premium of $18,000 has been paid in cash in advance.*

The prepayment of the annual premium of insurance is another Prepaid Expenses. Its nature is the same as that of Prepaid Rent introduced in Example 1.4 earlier. The journal entry is shown below.

Table 1-10(A) Example 1-06 Journal Entry

Debit			Credit		
Impact	Account	Amount	Account	Amount	Impact
A+	Prepaid Insurance	18,000	Cash	18,000	A–

R&B purchased protection by depositing the annual premium in advance. This deposit will be used for monthly insurance expenses during the next 12 months. The relevant *T*-accounts after posting and the updated Balance Sheet are presented below.

Table 1-10(B) Example 1-06 Posting

Cash			
Ex 1-01	450,000	100,000	Ex 1-04
Ex 1-02	150,000	250,000	Ex 1-05
Ex 1-03	200,000	18,000	Ex 1-06

Prepaid Insurance	
Ex. 1-06	18,000

Table 1-11 The Balance Sheet After Fire Insurance Premium (Ex. 1-06)

The Balance Sheet R&B Grill, Inc. On January ##, 2XX1 – after Fire Insurance Purchase			
Assets:		Liabilities & Equity	
		Liabilities	
Cash	432,000	Notes payable	400,000
Prepaid Rent	100,000		
Prepaid insurance	18,000	Equity	
		Common stock	
		– Rachel & Brad	450,000
FF&E	500,000		
		Common stock	
		– Jane	200,000
Total Assets	1,050,000	Total Liabilities & Equity	1,050,000

Let us assume R&B Grill is ready to start its business. To serve customers, Rachel and Brad must have inventory. Example 1.7 presents the purchase of inventory.

Example 1.7: *Inventory (F&B) of $80,000 was purchased. One half of the amount was paid in cash and the rest was put on account for 1 month. The journal entry of this transaction is presented below.*

Table 1-12(A) Example 1-07 Journal Entry

Debit			Credit		
Impact	**Account**	**Amount**	**Account**	**Amount**	**Impact**
A+	Inventory	80,000	Cash	25,000	A–
			Accounts Payable	55,000	L+

Purchases of inventories are considered as short-term investing activities of a business. This spending is not an expense. As shown in the journal entry in Table 1.12(A), this transaction has increased the company's Assets by $80,000 (in the "Inventory" account) and decreased its Assets by $25,000 (in the "Cash" account), and it also increased its Liabilities by $55,000 (in the "Account Payable" account). This transaction, in its nature, is exactly the same as previous examples of purchasing other asset items. When inventories are consumed over time, only the amount of the used inventory will be recorded as an expense. The following Table 1.12(B) presents the *T*-accounts when these records are posted. It must be noticed that the unpaid amount for the purchase is put into the "Accounts Payable." It is a Current Liabilities account that records the amount to be paid within 1 year. On the other hand, "Notes Payable" account belongs to the long-term Liabilities and is used for the amount to be paid for longer than 1 year. Following Table 1.13 shows the updated Balance Sheet after this transaction.

The balance sheet (in Table 1.13) presents a few items in the asset side with the total of $1,105,000. The liability section shows two different accounts – Accounts Payable and Notes Payable. All examples recorded so far have changed the financial position of R&B Grill as shown on the current balance sheet. However, R&B Grill has not even served a single

Table 1-12(B) Example 1-07 Posting

Cash				Inventory	
Ex 1-01	450,000	100,000	Ex 1-04	Ex 1-07	80,000
Ex 1-02	150,000	250,000	Ex 1-05		
Ex 1-03	200,000	18,000	Ex 1-06		
		25,000	Ex 1-07		

Accounts Payable			
		55,000	Ex 1-07

Table 1-13 The Balance Sheet After Example 1-07 Inventory Purchase

The Balance Sheet R&B Grill, Inc. On January ##, 2XX1 – after Inventory Purchase			
Assets:		Liabilities & Equity	
		Liabilities	
Cash	407,000	Accounts Payable	55,000
Inventory	80,000	Notes Payable	400,000
Prepaid Rent	100,000		
Prepaid Insurance	18,000	Equity	
		Common Stock	
		– Rachel & Brad	450,000
FF&E	500,000		
		Common Stock	
		– Jane	200,000
Total Assets	1,105,000	Total Liabilities & Equity	1,105,000

customer yet. With no customer served, there is no revenue. It is time to discuss the operational dimension of the business that brings in revenues and incur expenses.

1.10 REVENUES AND EXPENSES – OPERATING ACTIVITIES

As we have seen, R&B Grill has increased its asset volume from $800,000 on the first day to $1,105,000 as it is prepared for its operations. Although the amount of entire assets has increased, the amount of its owners' investment (equity) remained the same. It is important to understand that the only way to increase the value of the owner's investments is through generating Profits from the operations. Operations can be defined as the activities that involve serving customers. Serving customers brings in revenues, but it also incurs necessary expenses. The surplus or the leftover from these activities is called the profit of the business; and the result of these activities is described in an equation of Revenues – Expenses = Profit. When the total amount of the expenses is larger than that of the revenues, profit becomes negative and it is called Loss.

1.11 REVENUES

Up to this point, two major equations of accounting have introduced. One is the equation of the financial position presented as A = L + E. The other is the equation of the operating result presented as Revenues – Expenses = Profit. One frequent confusion about the revenues and expenses needs to be clarified. Revenues are often misunderstood as the cash received. Revenues are just the amount of services rendered with products delivered to customers, while cash is an asset item. A firm records its revenues when it serves its customer. In return, it will receive the fee charged in cash. This frequent misunderstanding comes from mixing the payment recorded (Cash or Accounts Receivable) with the activity of serving customers (Revenues). To help clarify this common misunderstanding and other similar confusion, this book has taken the journal entry format that emphasizes the impact on both sides (debit and credit).

1.12 EXPENSES

Another common confusion is about expenses. Similar to the revenues, expenses should be separately viewed from the spending of cash. As previously pointed out, spending cash for inventory purchase is not an expense. Likewise, spending for purchasing furniture or equipment is not an expense, either. Expenses are the amount of assets used (or consumed) in operating activities. For example, the amount of inventory used to serve customers is recorded as an expense. More specifically, the amount of inventory used in serving customers is recorded as the Cost of Sales. Besides, expenses also cover the amount of consuming the services or products provided by other people or businesses. A good example of this is the Payroll Expense. The amount of the services provided by the employees is recorded as Payroll Expenses (or Labor Expenses or under other terms). Another example is the Utility Expense that covers the amount of consumed products and services provided by power companies. Rent Expense, in this context, is the amount of using the space that belongs to the landlord. Interests are the expenses of using the services of the creditors. The payments of these must be recorded separately. When cash is paid for an expense, assets decrease (because cash is paid) for the expense. When the payment is delayed, the unpaid amount of the expense becomes a debt, and it increases the Liabilities of the business. In conclusion, Revenues and Expenses must be separately addressed from Assets or Liabilities accounts that are used to indicate the payment method.

Another important concept of expenses is that expenses represent the spending to support operating activities. When serving a customer, products must be delivered and necessary labor is performed. In addition, other supporting services are rendered. Necessary labor includes the preparation and the delivery of the product while other services rendered include the maintenance of the place, lighting, heating and air-conditioning, and repair of property, and many other similar activities. Let's continue with a few more examples of operating-related transactions.

Example 1.8: *A customer signed a contract for a party in 2 weeks with the guarantee of $15,000. A check of $8,000 has been received to prepare for the party. The rest will be paid when the party is over.*

This transaction requires careful attention to details. The accounting information only picks up the actual monetary value exchanged. For this reason, the guaranteed amount is not accounting information. It is just a promise. The real and the only information in this example to record is the cash amount received ($8,000). At the same time, it is not revenue because there has been no service rendered. Instead, the cash received holds R&B Grill responsible for the promised service later. By agreeing to provide the service in the future and accepting the deposit in cash, R&B Grill has imposed a liability on itself. The frequently used account title for this type of deposit is "Advanced deposit." Another title of "Unearned revenues" is also often used. We will use "Unearned revenues" because it explains the nature of the transaction more precisely. Because it is "unearned" at the time of recording, it is not revenue and the amount becomes a liability to the business because it owes services in the future. The journal entry and the posting of this transaction are presented below, followed by the Table 1-15 "the Balance Sheet After the Party Contract (Example 1-08) that reflects the impacts of this transaction on Assets and Liabilities.

Table 1-14(A) Example 1-08 Journal Entry

Debit			Credit		
Impact	Account	Amount	Account	Amount	Impact
A+	Cash	8,000	Unearned Revenues	8,000	L+

Table 1-14(B) Example 1-07 Posting

Cash				Unearned Revenues			
Ex 1-01	450,000	100,000	Ex 1-04			8,000	Ex 1-08
Ex 1-02	150,000	250,000	Ex 1-05				
Ex 1-03	200,000	18,000	Ex 1-06				
Ex 1-08	8,000	25,000	Ex 1-07				

Table 1-15 The Balance Sheet After Party Contract (Example 1-08)

The Balance Sheet R&B Grill, Inc. On January ##, 2XX1 – after a Party Contract			
Assets:		**Liabilities & Equity:**	
		Liabilities	
Cash	415,000	Accounts Payable	55,000
Inventory	80,000	Unearned Revenues	8,000
Prepaid Rent	100,000	Notes Payable	400,000
Prepaid Insurance	18,000		
		Equity	
		Common Stock	
		– Rachel & Brad	450,000
FF&E	500,000		
		Common Stock	
		– Jane	200,000
Total Assets	1,113,000	Total Liabilities & Equity	1,113,000

The cash balance on the balance sheet increased by $8,000 to $415,000, and the liabilities have also increased by $8,000. The bottom line of the balance sheet shows the increased amount in each group. As mentioned, this transaction has nothing to do with the revenues and/or expenses yet. The next transaction explains how this contract of future service is converted to revenues.

Example 1.9: *The customer mentioned in Example 1.8 hosted the party on the premise. The total amount for the party was $18,000. According to the contract, an invoice (bill) was handed for the balance outstanding.*

Now that the service is rendered as agreed, R&B Grill must record the amount as its revenues. Although the guaranteed amount of the service was $15,000 at the time of contract, the actual service turned out to be $18,000. The business should record revenues of $18,000. Revenues are recorded in the credit side because they technically increase owners' equity. This will be explained later in this section. With the services rendered, R&B Grill does not owe any more to the client for the deposit of $8,000 received earlier. This is the same as paying back the liability. The rest of the fee has not been collected in cash and a bill has been sent out for future collection. This uncollected fee is recorded in a new asset account called "Accounts Receivable."

Table 1-16(A) Example 1-09 Journal Entry

Debit			Credit		
Impact	**Account**	**Amount**	**Account**	**Amount**	**Impact**
A+	Accounts Receivable	10,000	Revenues	18,000	Rev/E+
L-	Unearned Revenues	8,000			

It must be noticed that the liability account of the "Unearned Revenues" is recorded in the debit side indicating it has decreased. Also, the debit side is recorded with the Account Receivable that indicates R&B Grill needs to collect the amount as soon as possible. The credit side recorded the revenues of the party service rendered. The impact of it is presented in two different ways – Revenues or Equity increase (+). Again, although impacts are not recorded in accounting practice, they are introduced in this book to help readers develop better understanding about the effect of accounting information. Posting is presented in the following.

Table 1-16(B) Example 1-09 Posting

	Accounts Receivable		
Ex 1-09	10,000		

	Unearned Revenues		
Ex 1-09	8,000	8,000	Ex 1-08

	Revenues		
		8,000	Ex 1-09

Revenues, as mentioned earlier, technically increase Owner's Equity. However, revenues are not directly put in to the Balance Sheet. Instead, revenues and expenses are separately summarized in the Income Statement to present the results of the operation. Only the final result, which is the profit or loss, is transferred to the Balance Sheet. For this reason, the Balance Sheet will not be presented until all transactions are completed. The *T*-accounts that belong to Assets, Liabilities, and Equity will continuously be presented as they change according to the transactions that follow.

The next example introduces the usual practice of reporting revenues in restaurant operations. Restaurants constantly serve customers and the cash collected is frequently deposited. However, the revenues are usually consolidated and recorded at the end of each day as daily revenues, or even once a week. The transaction below is an exaggerated example of this practice.

> **Example 1.10:** *During the rest of the month, R&B Grill has generated revenues of $97,000. The cash amount received from the sales was $47,000. All were deposited immediately. The rest of the sales were put on as account, and they have not been collected yet. This uncollected amount does not include the party service rendered in the previous example (Example 1.9).*

This example includes a few additional principles of accounting that need further explanation. All businesses are required to keep the hard evidence of their accounting records. It is called the principle of Object Evidence. In case of revenues, the most important evidence is, of course, Point-of-Sales (POS) receipts. Besides, there could be contracts, invoices, and many other supporting documents that support the accounting records. The same is true with all expenses.

Another new term used in this transaction is "on account." It means that the payment has not been made. In case of revenues, they were put "on account," the customer did not pay the fee. This becomes "Accounts Receivable" to the business, which is an asset item. On the other hand, if a purchase is made "on account," the buyer did not pay the price. So, it becomes the "Accounts Payable" that is a liability account to the business. The following is the journal entry and the *T*-accounts after posting of this transaction.

Table 1-17(A) Example 1-10 Journal Entry

Debit			Credit		
Impact	Account	Amount	Account	Amount	Impact
A+	Cash	47,000	Revenues	97,000	Rev/E+
A+	Accounts Receivable	50,000			

Table 1-17(B) *T*-Accounts After Revenues Recording (Example 1-10)

Cash			
Ex 1-01	450,000	100,000	Ex 1-04
Ex 1-02	150,000	250,000	Ex 1-05
Ex 1-03	200,000	18,000	Ex 1-06
Ex 1-08	8,000	25,000	Ex 1-07
Ex 1-10	47,000		

Revenues			
		18,000	Ex 1-09
		97,000	Ex 1-10

Accounts Receivable			
Ex 1-09	10,000		
Ex 1-10	50,000		

Now that R&B Grill has a couple of records of revenues, expenses should be discussed. As mentioned earlier, expenses are incurred simultaneously with the revenues when customers are served. Expenses can be broadly defined as the spending of the business that takes place during its operations. There are two different types of expenses. One is operating expenses, and the other is non-operating expenses. In this classification, the term "operating" specifically refers to the direct activities of serving customers. This chapter will focus on the "operating expenses" and the "non-operating expenses" will be explained in Chapter 3.

Serving customers includes the delivery of the product. It also involves supporting activities of labor such as preparing the product and delivery of it. These are considered as the spending directly related to serving customers. At the same time, the business must spend other expenses that are not directly related to serving customers. A good example is keeping the place clean and making it look attractive to customers. These include, but are not limited to, cleaning, maintenance, decorating, heating and air-conditioning, and so forth. All these activities cause perpetual spending to the business. The common factor of these expenses is that the business has consumed either its own assets – such as inventory – or other parties' services, as explained earlier.

No examples introduced so far have included expenses. Although most of them will be explained fully in the next phase, now it is a good time to introduce a couple of simple expenses. Let us review the following examples.

Example 1.11: *A Utility bill of $5,000 was received. It was put on account as the due date of the payment is in 2 weeks.*

This is rather an easy transaction. R&B Grill has consumed electricity, which was the product and services of the Power Company, in the amount of $5,000 during the month. Although the consumption of electricity is continuous, the business reports the expense when the bill is received. As the example says, the payment has been postponed. Regardless of the payment, however, the expense must be recognized as follows:

Table 1-18(A) Example 1-11 Journal Entry

Debit			Credit		
Impact	Account	Amount	Account	Amount	Impact
Exp	Utility	5,000	Accrued Expenses	5,000	L+

Table 1-18(B) Example 1-11 Posting

Utility Expense			Accrued expenses	
Ex 1-11	5,000		5,000	Ex 1-11

This transaction introduced a new *T*-account of "Accrued Expenses," which accommodates unpaid amount of expenses. Previously, unpaid amount of inventory purchases was put into the "Accounts Payable." The subtle difference is that the unpaid amount of purchases (usually for operations) uses the "Accounts Payable," while unpaid amounts of expenses are put into "Accrued Expenses." The "Accrued Expenses" is also a liability of the business. Let's record one more expense item that was incurred during the operations up to this point.

Example 1.12: *The monthly payroll amount for employees was $24,000. However, the payday is still 10 days ahead.*

This transaction is in regards to the payroll expense. Most businesses use the term "Labor Expense" for this purpose. R&B Grill apparently has conducted its operations and served customers. In return, it has recorded its revenues. As mentioned earlier, the activities of generating revenues automatically incur expenses by consuming the resources of the business, the labor of employees, and necessary services provided by other businesses. In this transaction, only the payroll expenses will be recorded. It must be noticed that the payment has not been made although the expenses have been recognized as practiced in the previous example. The journal entry looks like the following:

Table 1-19(A) Example 1-12 Journal Entry

Debit			Credit		
Impact	**Account**	**Amount**	**Account**	**Amount**	**Impact**
Exp	Labor Expenses	24,000	Accrued Expenses	24,000	L+

Table 1-19(B) Example 1-12 Posting

Labor Expenses			Accrued Expenses		
Ex 1-12	24,000			5,000	Ex 1-11
				24,000	Ex 1-12

The "Accrued Expenses" *T*-account now shows two records on the credit side – one for the unpaid expense of utility (Example 1.11) and the other for the unpaid expense of payroll (Example 1.12) – indicating that R&B Grill has increased its liabilities owing $29,000 to the parties that have provided their services. The creditors for these are the utility company and the employees of the business.

1.13 THE INCOME STATEMENT – TEMPORARY RESULT

When Example 1.8 was completed, a temporary balance sheet was introduced in Table 1.15 to show the financial position of R&B Grill at the time. Since then, only the journal entries and *T*-accounts of revenues and expenses were presented. It is because revenues and expenses are not reported on the Balance Sheet. Instead, they are separately presented in the Income Statement. For this reason, until the Income Statement is completed, the Balance Sheet cannot be prepared. In addition, as we saw in the transactions of Examples 1.9–1-12, revenues and expenses affect assets and liabilities. These examples were introduced to show that the operations of a business always change the firm's financial position. This is a very important dynamic of a business that must be understood. Before R&B Grill generated its first revenues, its financial position (after the Example 1.9) was presented in the Balance Sheet in Table 1.15.

Before going further to consolidate the operational results of the Example 1.9–1.12, let us summarize the financial status of R&B Grill when it was ready to serve customers. It had $1,113,000 in its assets. Of the total asset amount, it owed $463,000 (which is the total of the liabilities) to its creditors. Put differently, the creditors claim $463,000 on the R&B's total Assets. The rest of the Assets ($650,000) are claimed by the owners. The lists of assets, liabilities, and equity indicate the details of asset items the business owns, for what types of liabilities it owes, and who have invested as owners. The Balance Sheet, however, does not tell the amount of revenues generated or expenses incurred. Now, it is a good time to organize the operational result by summarizing the revenues and expenses, which also have changed the status of the assets and liabilities. When Assets and Liabilities are affected, the Equity must be affected due to the accounting equation of $A = L + E$.

The Balance Sheet After Party Contract (Example 1-08)

The Balance Sheet R&B Grill, Inc. On January ##, 2XX1 – after a Party Contract			
Assets:		**Liabilities & Equity:**	
		Liabilities	
Cash	415,000	Accounts Payable	55,000
Inventory	80,000	Unearned Revenues	8,000
Prepaid Rent	100,000	Notes Payable	400,000
Prepaid Insurance	18,000		
		Equity	
		Common Stock – Rachel & Brad	450,000
FF&E	500,000		
		Common Stock – Jane	200,000
Total Assets	1,113,000	Total Liabilities & Equity	1,113,000

The summary of the revenues and expenses so far is as follows: The first revenue record in the Example 1.9 (the transaction of party service, as presented in Table 1.13) recorded three entries: $18,000 of revenues in the credit side; $10,000 of Accounts Receivable in the debit side (indicating asset increase); and finally, $8,000 of Unearned Revenues in the debit side indicating the decrease of the liability. The next example (Example 1.10) recorded additional revenues of $97,000. Of this amount, $47,000 had been collected in cash and the rest ($50,000) has been put into "Accounts Receivable" for future collection. With these transactions, the total revenues became $115,000 and cash was increased by $47,000 and finally the total balance of the "Accounts Receivable" became $60,000. The next two transactions (Examples 1.11 and 1.12) reported two expense accounts. Because these expenses were not paid, the unpaid amounts were recorded in the "Accrued Expenses" as liabilities. The consolidated records with the balance of individual *T*-accounts are presented in the following Table 1.20.

Table 1-20 *T*-Accounts Balances for Intermediary Verification

Asset accounts

Cash

Ex 1-01	450,000	100,000	Ex 1-04
Ex 1-02	150,000	250,000	Ex 1-05
Ex 1-03	200,000	18,000	Ex 1-06
Ex 1-08	8,000	25,000	Ex 1-07
Ex 1-10	47,000	393,000	Total
Total	855,000		
Bal.	462,000		

Accounts receivable

Ex 1-09	10,000		
Ex 1-10	50,000		
Bal.	60,000		

Inventory

Ex 1-07	80,000		
Bal.	80,000		

Prepaid rent

Ex. 1-04	100,000		
Bal.	100,000		

Prepaid Insurance

Ex. 1-06	18,000		
Bal.	18,000		

FF&E

Ex 1-05	500,000		
Bal.	500,000		

Liability accounts

Accounts payable

		55,000	Ex 1-07
		55,000	Bal.

Accrued expenses

		5,000	Ex 1-11
		24,000	Ex 1-12
		29,000	Bal.

Unearned revenues

Ex 9	8,000	8,000	Ex 1-08
		–	Bal.

Notes payable

		150,000	Ex 1-02
		250,000	Ex 1-05
		400,000	Bal.

Table 1-20 (Continued)

Equity accounts

Common stock – Rachel & Brad			
		450,000	Ex 1-01
		450,000	Bal.

Common stock – Jane			
		200,000	Ex 1-03
		200,000	Bal.

Revenues & Expenses

Revenues			
		18,000	Ex 1-09
		97,000	Ex 1-10
		115,000	Bal.

Labor expense			
Ex 1-12	24,000		
Bal.	24,000		

Utility expense			
Ex 1-11	5,000		
Bal.	5,000		

Table 1.20 shows the balance of each *T*-account used in the recording process up to this point. It is vertically organized by the account group. The asset account group has Cash, Accounts Receivable, Inventory, Prepaid Rent, Prepaid Insurance, and FF&E. All these accounts have their balances on the debit side. The Cash account is the only one that has credit records showing the transactions that caused cash to decrease. The debit total (of Cash) of $855,000 is the total amount of cash the business has obtained from the beginning until now. The credit total shows the total amount of cash spent during the same time, which is $393,000. As a result, R&B Grill currently has $462,000 of cash on hand. All other asset accounts have only debit records and their balances are presented in the debit side.

The liability accounts, on the other hand, show credit records. As we know, the increase of liabilities is presented in the credit side. Their balances are also recorded in the credit side. The same is true with the Owners' Equity accounts. There are two separate Common Stock accounts that belong to the Owners' Equity group for two different groups of shareholders. All records, including the balances, are shown on the credit side.

Summarizing the nature of the *T*-accounts and their balances, a conclusion can be made at this point. The balance of individual *T*-accounts is recorded in the "increasing" side. As asset accounts increase in the debit, their balances are presented in the debit side. Liability and Equity accounts, in contrast, record their balances on the credit side because their increases are recorded on the credit side. As we proceed with more advanced concepts and techniques, a few exceptions will be introduced. This simple rule is usually introduced as "normal balance" in accounting textbooks.

Interestingly, there is at least one asset account that records its balance in the credit side. This will be introduced in the next chapter with an example of Accumulated Depreciation. However, even in such an exceptional situation, the rule of balance recording in the increasing side remains the same. Even in such occasions, the meaning of the credit records of an asset account – decrease – does not change. The credit record of an asset account indicates the impact of "decreasing." Therefore, the balance of an asset account in the credit side represents a negative value.

Let's prepare the Balance Sheet by transferring the balances of each account to appropriate groups. Because revenues and expenses, with their impacts on Assets, Liabilities, and Equity, are reported separately on the Income Statement, the equation of Assets = Liabilities + Equity does not seem to work now. Take a look at the following Balance Sheet newly created in the Table 1.21. Pay attention to the bottom line for the amount of the total assets and the total liabilities + equity.

Table 1-21 The Balance Sheet for Temporarily Closing (Incomplete)

The Balance Sheet The R&B Grill, Inc. On Jan. XX, 2XX1			
Assets		**Liabilities & Equity**	
		Liabilities	
Cash	462,000	Accounts Payable	55,000
Accounts receivable	60,000	Accrued Expenses	29,000
Inventory	80,000	Unearned Revenues	-
Prepaid Rent	100,000		
Prepaid Insurance	18,000	Notes Payable	400,000
		Total Liabilities	484,000
FF&E	500,000	Owners' Equity	
		Common Stock – Rachel and Brad	450,000
		Common Stock – Jane	200,000
		Total Owners' Equity	650,000
Total Assets	1,220,000	Total liabilities & Equity	1,134,000

* At this point, the Balance Sheet shows different amounts of Assets and Liabilities and Equity combined. The difference is Assets $1,220,000 – L+E $1,134,000 = $86,000. (Review the Income Statement for this.)

The bottom line of the Balance Sheet (Table 1.21) shows that the equation of $A = L + E$ is not working. The total asset amount is larger than the liabilities and equity combined by $86,000. The reason is, as mentioned before, that the effects of revenues and expenses have not been included in the Balance Sheet as of yet. Only with the operational results transferred from the Income Statement, the Balance Sheet can be completed. The following Table 1.22 shows the summarized records of revenues and expenses up to now.

Table 1-22 The Temporary Income Statement
(refer to the comments on the bottom)

The Income Statement The R&B Grill, Inc. For the month of January, 2XX1		
Revenues		115,000
Expenses		
Labor expense	24,000	
Utility expense	5,000	
Total expenses		29,000
Profit (Loss)		86,000

* This Income Statement shows only the records that have been covered so far.

* More expenses that have been incurred during the operations will be introduced in the next chapter to complete the Income Statement.

The bottom line on the Income Statement shows a Profit of $86,000, which is exactly the same as the overage of the assets against the total liabilities and equity combined (as shown in the Table 1.21). As revenues increase owners' equity and expenses decrease owners' equity, profits will eventually increase the equity. In case the business suffered losses, it would decrease the equity. This is why revenues and expenses are not directly reported on the Balance Sheet and they are called temporary accounts of the owners' equity.

The term Profit or Loss does not belong to the balance sheet because it is the operational results. It has to transform its nature of the operating result to that of the equity when transferred to the Balance Sheet. At this transformational step, it becomes a part of the "Retained Earnings" that belongs to the owners' equity. The term "Retained Earnings" indicates that the earnings from the operations have been kept by the company for the owners. When the profit reported in the Income Statement is added to the section of the Owners' Equity on the balance sheet under the account title of "Retained Earnings," the discrepancy between Total Assets (A) and Liabilities and equity combined ($L + E$) disappears and the accounting equation of $A = L + E$ works again as shown in Table 1.23.

Table 1-23 Revised Balance Sheet for Temporarily Closing

The Balance Sheet The R&B, Inc. On Jan. XX, 2XX1			
Assets		**Liabilities & Equity**	
		Liabilities	
Cash	462,000	Accounts payable	55,000
Accounts receivable	60,000	Accrued expenses	29,000
Inventory	80,000	Unearned revenues	–
Prepaid rent	100,000		
Prepaid insurance	18,000	Notes payable	400,000
		Total Liabilities	484,000
FF&E	500,000	Owners' Equity	
		Common stock – Rachel and Brad	450,000
		Common stock – Jane	200,000
		Retained Earnings	86,000
		Total Owners' Equity	736,000
Total Assets	1,220,000	Total liabilities & Equity	1,220,000

* Now with the Profit of $86,000 added into the Retained Earnings,
The equation of A = L + E works out.

There is one more fact to keep in mind regarding Retained Earnings. As profits (or losses) from operations are continuously retained by the business for its owners, it accumulates whenever new profits (or losses) are reported. In this context, the Retained Earnings can be defined as "accumulated profits (or losses) throughout the life of a business." This is how the operations of a business contribute to the growth of owners' investment.

When losses are reported, the amount of Retained Earnings is reduced. Simply, it reduces the value of the owners' equity. For this reason, the Retained Earnings may become negative on the Balance Sheet.

If a business continuously reports losses from its operations for long, the accumulated negative amount of its Retained Earnings may bring down the entire value of its owners' equity into a negative amount.

1.14 CONCLUSION

This chapter has introduced the general concept of accounting with several examples of common business transactions. It showed how business activities of financing, investing, and operations are organized through the recording process and developed into meaningful business information. However, the examples introduced in this chapter are too limited to conclude the business situation and activities yet. One example of the missing information is the spending of the products served to the customer. It is easy to imagine that serving customer involves the delivery of products. This information is completely missing in this chapter. Another is the situation of the Accounts Receivable and Accounts Payable. What if the business has collected its outstanding fees from those who had not paid immediately? Or, what if the business has paid its debts it had put on hold? What happens to the value of assets such as equipment and furniture? Do they maintain their original value regardless of their usage? What happens to the expenses that were paid in advance? Has not the business spent some of its prepaid expenses? These details must be investigated and reported accurately. Usually, these are introduced as "Adjusting Entries" in most accounting textbooks. They often introduce a template for this process. This book, however, explains these as an extended process of journalizing and posting. This will be covered in the next chapter.

CHAPTER 2

ADJUSTING ENTRIES FOR MISSING INFORMATION

2.1 INTRODUCTION

The previous chapter introduced how business activities are transformed into meaningful accounting information. However, it was incomplete. Accounting provides structured methods of organizing raw data of business transactions, but there are multiple steps of refining information to add more meaning. Those steps are known as the recording process.

In the recording process, each transaction is compressed into relevant account titles and amounts. Then, it is further processed to determine the increase or decrease of Assets, Liabilities, and Equity. Once these are fully determined, each record is transferred to the correct side of debit or credit. In this process, the increase of assets of the company was put in the debit side, and the increase of liabilities and/or equity is put in the credit side. The decrease of assets, liabilities, and/or equity takes the opposite side. One important rule in this process is that when an entry is made on one side, there must be at least one other entry put on the other side. The total amount of both sides must be identical for every single transaction and for all transactions as well. The following is the replica of the rules of journal entry that summarize this concept presented earlier in Chapter 1.

The dynamics of dual entry system introduced in the Figure 1.1 does not show revenues or expenses. As pointed out in Chapter 1, revenues increase owners' equity and expenses decrease owners' equity. However, they are not directly recorded into the owners' equity section of the Balance Sheet. Instead, they are separately recorded and then summarized into a separate report to show the operational result of the business. These concepts – revenues and expenses – must be added to the diagram with the following logic.

Revenues increase equity by causing gains and expenses decrease equity by causing losses. For this reason, eventually, it is either gains or losses that affect the equity of a business. Although it has been emphasized earlier that revenues simultaneously incur expenses, the recording process clearly distinguishes the timing of the revenues and expenses and record them separately.

The following diagram (Figure 2.1) is an updated one that includes the concepts of gain and loss. When a business achieves gains, its assets will increase. On the other hand, when it suffers losses, its assets will decrease. Gains or losses may also affect the liabilities. It must be noticed that the diagram does not introduce direct connections between gains and

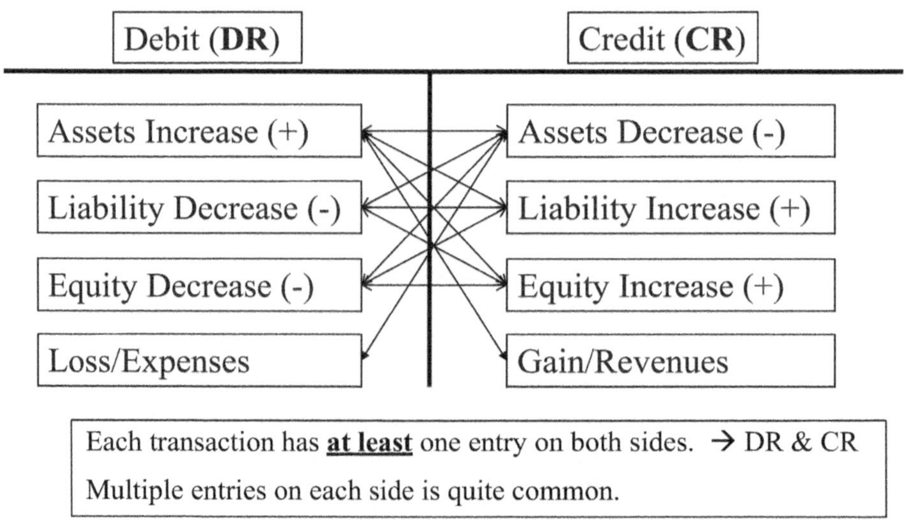

Figure 2.1 Double Entry System: Rules of Journal Entry Debit & Credit with Assets, Liabilities, Equity, Gain & Loss.

losses; and equity is not connected directly with gains or losses due to the timing issue of the recording process as explained above.

2.2 ADJUSTING ENTRIES TO UNCOVER HIDDEN ACTIVITIES

This chapter introduces many other expenses that were not presented in the previous chapter. The closing procedure that produced the tentative Balance Sheet at the end of Chapter 1 was a temporary step taken to verify the accuracy of all records up to that point. With the technology available today, few businesses conduct such time-consuming activities. Modern computerized system of financial accounting only requires the initial entry of journalizing correctly with predetermined account numbers. Then, all mechanical procedures are automatically conducted. However, even the most advanced computerized system still takes the conceptual steps of putting them into correct sides as introduced so far. Business practitioners must understand how the old-fashioned manual mechanism works to make sound decisions in their business. For this purpose, the goal of this book is not just about teaching how to conduct financial accounting procedures. Instead, it goes far beyond to help readers develop comprehensive knowledge about how to use accounting as their effective tool.

Although most expenses are recognized when bills arrive as we saw in the previous chapter, there are other expenses that need to be calculated or even to be estimated with no supporting documents. As pointed out, there must be more expenses than were introduced in Chapter 1. Let's start with the spending of the inventory-serving customers.

2.2.1 Cost of Sales

When inventory was purchased, it was recorded as the increase of an asset. The account title used was "Inventory." The payment made in cash has decreased assets. The unpaid amount was put into Account Payable and it increased the liability of the business. There were no expenses involved in the purchase. After one accounting period (e.g., a month,) is completed, the business must conduct a physical inventory by counting everything remained. Otherwise, there is no knowing of the amount of inventory consumed. The amount of used inventory is measured by applying the formula of "Beginning Inventory + all purchases – Ending Inventory." The result, which is the consumed amount of inventory, is called the "Cost of Sales" or the "Cost of Goods Sold."

Beginning Inventory + Purchases – Ending Inventory = Cost of Sales

Example 2.1: *Rachel conducted the monthly physical inventory counting and found the total amount of remaining inventory was $34,800.*

This transaction is about calculating the used amount of inventory, which is the Cost of Sales. As R&B Grill started as a brand new business, it did not have beginning inventory in the beginning. The calculation of the Cost of Sales is as follows:

Beginning Inventory + Purchases – Ending Inventory = Cost of Sales

$$\$0 + \$80,000 - \$34,800 = \$45,200$$

The purchase amount of $80,000 was obtained from the debit total of the inventory *T*-account. The Cost of Sales of $45,200 indicates that R&B Grill has consumed $45,200 of Inventory to generate the total revenues of $115,000 as shown in the temporary Income Statement. The Cost of Sales is 39.3% of the revenue (Cost of Sales $45,200/Revenues $115,000 × 100 = 39.3%). Put differently, the company has spent 39.3 cents of its inventory to earn $1 of revenue. This result must be journalized and posted using the same method.

Table 2-01(A) Example 2-01-a Journal Entry

Debit			Credit		
Impact	**Account**	**Amount**	**Account**	**Amount**	**Impact**
Exp.	Cost of Sales	45,200	Inventory	45,200	A-

Table 2-01(B) Example 2-01 Posting

Cost of Sales			Inventory		
Ex 2-01	45,200	Ex 1-07	80,000	45,200	Ex 2-01

A new expense *T*-account for the Cost of Sales is created. As explained, the expense (Cost of Sales) is recorded in the debit side. In the Inventory *T*-account, the credit side is used to record the amount of inventory used, indicating that it has decreased by the same amount.

It is very important in a restaurant business to control its inventory tightly. If inventories are wasted, the Cost of Sales will become very high. This will, in turn, lower the profits. Or it may even cause losses from operations. All restaurateurs must have their target Cost of Sales in percentage to control their inventory usage. This will be explained further in the second part of this book.

The Inventory *T*-account now shows a credit record of $45,200 (Example 2.1). Its debit side still shows the original purchase of $80,000. Now that $45,200 has decreased, the balance is $34,800. This is the amount of the physical inventory conducted by Rachel as stated in the Example 2.1. The remaining inventory is the ending inventory of the current. The business enters the next month with this amount as its beginning inventory. To this, all purchases during the new month will be added and the new ending inventory amount will be subtracted to determine the Cost of Sales of the month. This procedure is called the "Periodic Inventory System," which requires physical counting to determine the actual amount of remaining inventory at the end of each accounting period.

2.2.2 Prepaid Expenses

The next expense item is the Rent. In Example 1.3, R&B Grill deposited $100,000 for its future rent expenses. It was recorded as "Prepaid Rent," which shows its balance of $100,000 in the debit side. The debit balance of an asset account indicates the amount is available for use. After 1 month, R&B Grill must report its expense for using the landlord's service.

Example 2.2: *R&B Grill recorded its monthly Rent Expense out of its deposited funds. The journal entry looks like the following:*

Table 2-02(A) Example 2-02-a Journal Entry

Debit			Credit		
Impact	Account	Amount	Account	Amount	Impact
Exp.	Rent Expense	15,000	Prepaid Rent	15,000	A-

Table 2-02(B) Example 2-02 Posting

Prepaid Rent			Rent Expense		
Ex. 1-04	100,000	15,000	Ex 2-02	Ex 2-02	15,000

According to the rules of journal entry, expenses are recorded in the debit side. When an expense is incurred, the amount can be paid in cash (which decreases assets). Quite often, the payment is postponed when the due date is far ahead. When the payment of an expense is deferred, the unpaid amount becomes a liability, and it is put into the "Accrued Expenses" account. However, this Example (2.2) is slightly different. It still decreases the company's assets but it does not spend cash. Instead, R&B Grill has used its deposited funds (stored in the "Prepaid Rent" account) by the amount of 1 month's rent. The "Prepaid Rent" *T*-account now has a new record in the credit side of $15,000, which shows that the amount has been spent. If the balance is calculated now, the "Prepaid Rent" would have a balance of $85,000. There are many occasions in a business where deposits must be made in advance, and expenses are taken away from the prepaid expenses when they are incurred. The subscription of magazines or newspaper is one example and insurance is another.

The next example is the case of the monthly insurance expense. The Example 1.6 in Chapter 1 introduced the annual premium of insurance prepaid. At the time, the *T*-account of the "Prepaid Insurance" was presented as following:

Example 1-06 Journal Entry

Debit			Credit		
Impact	Account	Amount	Account	Amount	Impact
A+	Prepaid Insurance	18,000	Cash	18,000	A–

Example 2.3: *R&B Grill has recorded the monthly insurance expense.*

Table 2-03(A) Example 2-03-a Journal Entry

Debit			Credit		
Impact	Account	Amount	Account	Amount	Impact
Exp.	Insurance Expense	1,500	Prepaid Insurance	1,500	A-

Table 2-03(B) Example 2-03 Posting

Prepaid Insurance				Insurance Expense			
Ex. 1-06	18,000	1,500	Ex 2-03	Ex 2-03	1,500		

Just like the "Prepaid Rent" account illustrated in the Example 2.2 (Tables 2.2(A) and (B)), this transaction has adjusted the "Prepaid Insurance" account by the monthly insurance expense amount (annual premium of \$18,000/12 = \$1,500 per month). The insurance expense was incurred because R&B Grill has used up the protection coverage of the insurance for 1 month, leaving the balance for the next 11 months' insurance premium still available in the deposited account. The "Prepaid Insurance" account presented in Table 2.6(B) shows the amount spent in the credit side, and the balance of the prepaid amount can be calculated by subtracting the credit total from the debit total.

2.2.3 Accrued Expenses

2.2.3.1 *Additional Labor Expenses (Payroll Expenses)*

Example 2.4: *Toward the end of Chapter 1, the initial labor expenses of \$24,000 were recorded without payment. The unpaid amount was recorded in the Accrued Expenses account. Payment was made correctly on the pay-day (i). It was also found that additional payroll expenses of \$14,000 were incurred. However, the payment of this new expense has not been made because the payday is still more than a week ahead (ii).*

This transaction includes two different business activities as titled "2.4-a-i" and "2.4-a-ii" in the following journal entry. Each activity is explained below the journal entry.

Table 2-04(A) Example 2-04 Journal Entry

Trx #	Debit Impact	Account	Amount	Credit Account	Amount	Impact
Ex. 2-04-a-i	L-	Accrued Expenses	24,000	Cash	24,000	A-
Ex. 2-04-a-ii	Exp.	Labor Expenses	14,000	Accrued Expenses	14,000	L+

The first activity ("2.4-a-i" in the journal entries) is about the payment of the liability recorded in the "Accrued Expenses" account (in Table 1.19 "Example 1.12"). The second one ("2.4-a-ii") is the additional labor expense incurred during the first month. As the payment of this new expense has not been made yet, the unpaid amount is recorded in the liability account of "Accrued Expenses." The following *T*-accounts explain the multiple impacts of this transaction.

Table 2-04(B) Example 2-04 Posting

Labor Expenses			
Ex 1-12	24,000		
Ex 2-04-a-ii	**14,000**		

Accrued expenses			
Ex 2-04-i	**24,000**	5,000	Ex 1-11
		24,000	Ex 1-12
		14,000	**Ex 2-04-a-ii**

Cash			
Ex 1-01	450,000	100,000	Ex 1-04
Ex 1-02	150,000	250,000	Ex 1-05
Ex 1-03	200,000	18,000	Ex 1-06
Ex 1-08	8,000	25,000	Ex 1-07
Ex 1-10	**125,000**	**24,000**	**Ex 2-04-a-i**

Table 2.4(B) shows the three *T*-accounts that were affected. First, the payment of the previously accrued labor expense (Example 1.12) is paid now (Example 2.4-(A-i)). The "Accrued Expenses" *T*-account is updated with a debit entry indicating the decrease of liabilities with $24,000. On the "Cash" *T*-account, the same amount ($24,000) is recorded in the credit side, which indicates the decrease of assets. Additionally, the newly incurred labor expense has been added as a new expense in the "Labor Expenses" *T*-account (Example 2.4-ii) in the debit side. On the credit side of the "Accrued Expenses" *T*-account, a new credit entry is added to show a new liability of $14,000 for the unpaid amount of expenses (Example 2.4-(ii)).

The "Accrued Expenses" *T*-account shows that the first entry (of Example 1.11, unpaid Utility Expense) has not been cleared yet. It is the utility expense of the first month that has not been paid. Readers may wonder why this has not been paid in practice. This example is provided to show that accounting records are processed only with documented record. Differently put, all records must have their supporting documents, such as invoices, canceled checks, internal reports (e.g. physical inventory list), or even hand-written memos. This is an application of the "Objective Evidence." For this principle, the unpaid utility expense still remains in the "Accrued Expenses" account. It is the responsibility of the management to solve this either by paying this or finding a record of the payment.

Another expense that was incurred during the first month is "Interest." The Example 1.2 introduced a transaction of a loan ($150,000). The journal entry only recorded the amount borrowed (principal). At the time, the interest expense on the loan was not reported. Now that 1 month has passed since the loan, R&B Grill needs to prepare its first set of official financial statements. The business must present every single expense on its Income Statement. The Example 2.6 provides the details of this loan. Please be advised that the conditions included in the following example are hypothetical.

2.2.3.2 Interest Expenses

Example 2.5: The loan of $150,000 (mentioned in the Example 1.2) was taken with the interest at 12% per year. Although the principal must be paid along with monthly interest, it is necessary at this point only to determine the amount of monthly interest expense to be recorded. More about the payment of long-term debt will be covered in Chapter 9. The monthly interest amount is calculated as follows:

$$\text{The annual interest amount} = \text{The balance of the loan} \times \text{Annual interest rate}$$

$$\text{The monthly interest amount} = \text{The annual interest} \times 1/12$$

Applying the formula presented above, the monthly interest expense is $1,500. This expense must be recorded, although it may not be paid immediately. This practice will be recorded as follows in the journalizing process.

Table 2-05(A) Example 2-05 Journal Entry

Debit			Credit		
Impact	Account	Amount	Account	Amount	Impact
Exp.	Interest Expense	1,500	Accrued Expenses	1,500	L+

This transaction increased the company's liability by deferring the payment of the interest expense to the creditor. As we practiced, the unpaid expense amount is put into the account of "Accrued Expenses," which is a liability account.

Table 2-05(B) Example 2-05 Posting

Interest Expense			
Ex 2-05-a	1,500		

Accrued Expenses			
Ex 2-04-a-i	24,000	5,000	Ex 1-11
		24,000	Ex 1-12
		50,000	Ex 2-04-a-ii
		1,500	Ex 2-05-a

With this recording, R&B Grill has reported its interest expense of $1,500 along with the increased liability of the unpaid amount of it. The "Accrued Expense" *T*-account shows the record on the bottom of its credit side (refer Table 2.5(B)). It is easily noticeable that the business has accumulated its liabilities by not paying a few expense items during the first month.

2.2.4 Depreciation

The next expense item that is incurred with no trace of records is Depreciation. Depreciation is the estimated amount of disappearing value of long-term assets, such as equipment, furniture, and other tangible items. A vehicle is an easy example to explain the concept of "Depreciation Expense." A vehicle is typical equipment that is used in a business for a long period. When it is purchased, the entire purchase price is recorded as an asset item of the business. The purchase amount is not an expense. Instead, it is considered as an internal investment of the company (resource allocation), for its operations as has been explained earlier. Once the vehicle is purchased, however, it starts losing its value over time, and the remaining value shrinks. The business must record the disappearing value as an expense periodically. This concept is applied to all other long-term assets. R&B Grill must record the depreciation of its FF&E. When R&B Grill has obtained its FF&E for $500,000 in the Example 1.5, the journal entry was created as follows:

Table 1-08(A) Example 1-05 Journal Entry

Debit			Credit		
Impact	Account	Amount	Account	Amount	Impact
A+	FF&E	500,000	Cash	250,000	A–
			Notes Payable	250,000	L+

The journal entry above clearly shows that the entire amount spent was put into the asset account of "FF&E." The result of this was also shown in the FF&E *T*-account. In other words, the company owns a new set of FF&E with the spending of the $500,000. This is considered as a long-term asset because it will be in use for longer than a year and it starts losing its value gradually. It is the responsibility of management to estimate the amount of the depreciation expense. There are three different methods of estimating the expense and each method has its own advantages and disadvantage as. In this section, the most commonly used technique – straight line method – is introduced. Once it is fully explained, the other two methods will be explained later with their advantages and disadvantages.

First, the management must determine how many years it plans to use the FF&E (or other long-term asset item). This is considered the "useful life" of the item. Then, the value of the item at the end of the useful life must be determined. This is called the "residual value" or "salvage value." These – useful life and residual value – are arbitrary decisions with no fixed rules. Once these two pieces of information are determined, then a fixed amount will be reported as the depreciation expense until the value of the FF&E reaches the predetermined amount of the salvage value. Let's assume the following additional conditions of R&B Grill's FF&E purchase.

Example 2.6: *Furniture, Fixture, and Equipment (FF&E) depreciation expense:*

When the business purchased FF&E, the vendor has reportedly said that many other businesses use the same type of FF&E for 5 years and trade it at around $140,000 for new sets.

These conditions provide the useful life of 5 years, and the residual value of $140,000. This helps Rachel and Brad to determine the reasonable amount of annual depreciation expense. As the residual value is set at $140,000, the total amount to be depreciated (for 5 years) will be $360,000 (the original price paid: $500,000 – residual value that can be traded in for the next purchase: $140,000 = the actual amount spent on the FF&E: $360,000). As the entire amount of $360,000 will have disappeared during the next 5 years, the annual amount will be $72,000 ($360,000/5 years = $72,000/year). When this annual depreciation is divided by 12 months, the monthly depreciation expense becomes $6,000. R&B Grill will incur additional expense of depreciation on the FF&E by $6,000 per month. Once the monthly depreciation expense is estimated this way, its journal entry looks like the following:

Table 2-06(A) Example 2-06 Journal Entry

Debit			Credit		
Impact	Account	Amount	Account	Amount	Impact
Exp.	Depreciation	6,000	Accumulated Depreciation	6,000	A-

The journal entry of depreciation introduced a unique account title of "Accumulated Depreciation." As its impact shows, it represents the decrease of assets. The "Depreciation," as an expense account, must be recorded on the debit side and on the credit side the account title to be recorded is always "Accumulated Depreciation (on FF&E)." "Accumulated Depreciation" is called a "Contra Asset," which means that it offsets the existing value of the asset item. Naturally, its value is already negative. This negative value justifies the credit entry as an asset account. It must be remembered that its balance is also presented on the credit side (A-). This is an exception to the normal balance location of assets. However, it was also pointed out that the balance always is recorded in the increasing side. In summary, as this account increases (its negative values offset the existing value of an asset item) in the credit side, its balance is also recorded in the same "credit side." When posting is completed with this journal entry, the *T*-accounts will look like Table 2-06(b).

Table 2-06(B) Example 2-06 Posting

Depreciation Expense – FF&E			
Ex 2-06	6,000		

Accumulated Depreciation – FF&E			
		6,000	Ex 2-06

The examples introduced in this chapter were of different nature compared with those introduced in the previous chapter. The common nature of the transactions of this chapter is that most of them have to be calculated or estimated by collecting a few different pieces of information. The Cost of Sales was calculated by applying the inventory information of beginning amount, purchases, and the result of the physical inventory. Prepaid expenses were adjusted for the time passage and labor expenses were adjusted with newly added amount. Interest expenses were calculated based on the long-term debt balance. Finally, depreciation expense on the FF&E was estimated using the business decisions on the residual value and useful life. Once all these additional records are presented, the next step is to finalize the result and prepare necessary reports.

2.3 BALANCE CALCULATION AND ERROR CHECKING WITH THE ADJUSTED TRIAL BALANCE

Most books on accounting separate the recording process of expenses into two parts: one for the simple ones as presented in Chapter 1 and the other under the title of the Adjusted Entry process. In between these two processes, as was presented in Chapter 1, an error-checking process called "the Trial Balance" is usually introduced. This book argues that the process of the trial balance can be ignored in practice thanks to the application of the computerized technology. However, it is always recommended to verify the accuracy of any work, particularly that with numerical information. For this reason, another step of error-checking procedure is introduced.

The following is the itemized list of traditional recording process of other books compared with the consolidated process used in this book.

A: Traditional Recording Process	B: Consolidated Process
1. Journal Entry	1. Journal Entry
2. Posting	2. Posting (no balance calculating)
a. Balance calculation	3. Adjusted Entry
3. Trial Balance to check error	a. Journalizing
4. Adjusted Entry	b. Posting
a. Journalizing	c. Balance Calculation
b. Posting	4. (Adjusted) Trial Balance to check error
c. Balance calculation	5. Financial statements
5. Adjusted Trial Balance to check error	
6. Financial statements	

As indicated above, the approach taken in this book ignores the intermediary step of the Trial Balance. By doing so, it has eliminated the step of the balance calculation for error checking at that point. With the advanced technology applications, the error-checking process may be obsolete. The step of the Adjusted Entry is just a redundant process of the journal entries and posting. As long as all transactions are recorded accurately into the journal records, there may be no need to take the extra step of Trial Balance. The only reason of the Adjusted Trial Balance presented in this book is to show how to check possible errors in the process up to this point.

Table 2-07 T-Accounts Balances for Closing the First Month

Asset accounts

Cash

Ex 1-01	450,000	100,000	Ex 1-04
Ex 1-02	150,000	250,000	Ex 1-05
Ex 1-03	200,000	18,000	Ex 1-06
Ex 1-08	8,000	25,000	Ex 1-07
Ex 1-10	47,000	24,000	Ex 2-04-a-i
	855,000	417,000	
Balance	438,000		

Accounts Receivable

Ex 1-09	10,000		
Ex 1-10	50,000		
Balance	60,000		

Liability & Equity accounts

Accounts Payable

		55,000	Ex 1-07
		55,000	Balance

Accrued Expenses

Ex 2-04-i	24,000	5,000	Ex 1-11
		24,000	Ex 1-12
		14,000	Ex 2-04-a-ii
		1,500	Ex 2-05
	24,000	44,500	
		20,500	Balance

Unearned Revenues

Ex 1-09	8,000	8,000	Ex 1-08
		–	Balance

Revenues & Expenses

Revenues

		18,000	Ex 1-09
		97,000	Ex 1-10
		115,000	Balance

Labor expense

Ex 1-12	24,000		
Ex 2-04-a-ii	14,000		
Balance	38,000		

Utility expense

Ex 1-11	5,000		
Balance	5,000		

Table 2-07 (Continued)

Inventory

Ex 1-07	80,000	45,200	Ex 2-01
Balance	34,800		

Prepaid Rent

Ex. 1-04	100,000	15,000	Ex 2-02
Balance	85,000		

Prepaid Insurance

Ex. 1-06	18,000	1,500	Ex 2-03
Balance	16,500		

Notes Payable

		150,000	Ex 1-02
		250,000	Ex 1-05
		400,000	Balance

Common Stock – Rachel & Brad

		450,000	Ex 1-01
		450,000	Balance

Cost of sales

Ex 2-01	45,200		
Balance	45,200		

Rent Expense

Ex 2-02	15,000		
Balance	15,000		

Insurance expense

Ex 2-03	1,500		
Balance	1,500		

FF&E

Ex 1-05	500,000		
Balance	500,000		

Common Stock – Jane

		200,000	Ex 1-03
		200,000	Balance

Interest expense

Ex 2-05	1,500		
Balance	1,500		

Accumulated Depreciation – FF&E

		6,000	Ex 2-06
		6,000	Balance

Depreciation – FF&E

Ex 2-06	6,000		
Balance	6,000		

2.3.1 Adjusted Trial Balance

The Trial Balance or the Adjusted Trial Balance is simply the listing of all *T*-account balances in each side followed by the total of each side on the bottom. The debit total must be identical to the credit total at this step. If they are not identical, there must be at least one or more errors in the *T*-accounts, or it may have originated from the journal entry. Any errors must be identified and corrected. One caution, however, is that even if the totals of both sides are the same, it is still possible that there could be an error if one entire transaction is not recorded at all. Taking the step of Trial Balance (or Adjusted Trial Balance) cannot find the error of the missing record. The conclusion of this is that the error-checking process of Trial Balance (or Adjusted Trial Balance) can only find the mechanical error of calculations. It is never capable of finding logical errors.

Table 2.7 "T-Account Balances for Closing the First Month" shows 21 *T*-accounts: there are seven asset accounts, four liability accounts, and two equity accounts. There are additional eight accounts that belong to the revenues and expenses group. As pointed out earlier, the balance of each *T*-account is recorded in the increasing side. Under this rule, all asset *T*-accounts show their balances on the debit side except the Accumulated Depreciation account. All *T*-accounts that belong to liabilities and equity show their balances on the credit side. Revenues account, as they increase owners' equity, presents the balance on the credit side, while expenses do their balances on the debit side because they decrease equity. The itemized list is presented with balances in the Adjusted Trial Balance in the Table 2.8. The bottom line of the table shows the same totals on both sides, and it is OK to proceed to prepare the Income Statement and the Balance Sheet. Again, the total amount on the Adjusted Trial Balance carries no other meaning but the records have been calculated correctly. It does not even mean that the all necessary records were put in the accounting book, but it is assumed that everything has been recorded correctly.

Once this step is successfully completed, the list in the Adjusted Trial Balance is used to prepare the Income Statement and the Balance Sheet. The Adjusted Trial Balance presents the account titles in the sequence of assets, liabilities, equity, revenues and expenses. From the list, the records of the revenues and expenses are used to prepare the Income Statement with the amount of profit or loss. This process is identical to the previous one introduced in Chapter 1.

Table 2-08 Adjusted Trial Balance – A

Adjusted Trial Balance – Before Income Tax		
Accounts	**Debit**	**Credit**
Cash	438,000	
Accounts Receivable	60,000	
Inventory	34,800	
Prepaid Rent	85,000	
Prepaid Insurance	16,500	
FF&E	500,000	
Accumulated Depreciation		6,000
Accounts Payable		55,000
Accrued Expenses		20,500
Unearned Revenues		–
Notes Payable		400,000
Common Stock – Rachel & Brad		450,000
Common Stock – Jane		200,000
Revenues		115,000
Cost of Sales	45,200	
Labor Expense	38,000	
Rent Expense	15,000	
Utility Expense	5,000	
Insurance Expense	1,500	
Depreciation Expense	6,000	
Interest Expense	1,500	
Total	1,246,500	1,246,500

2.4 INCOME TAX

During the recording process or the Adjusted Trial Balance, the Income Tax has not been mentioned. It is necessary to explain this last expense item now. When the Income Statement is prepared, the Income Tax must be assessed. When all expenses, including the Interest Expense, are subtracted from the total revenues, the remaining amount becomes the taxable income. A fixed tax rate is applied to this amount to calculate the Income Tax. In our practice, the tax rate of 35% is applied. The taxable income of R&B Grill in the first month is $2,800; and the amount of the Income Tax is 35% of it ($980). When the Income Tax is calculated, it becomes an additional expense but the payment is deferred until the Income Tax is due. In result, it is added to the Accrued Expenses. The journal entry will look like the following:

Table 2-09(A) Example 2-07 Journal Entry

Debit			Credit		
Impact	**Account**	**Amount**	**Account**	**Amount**	**Impact**
Exp.	Income Tax	980	Accrued Expenses	980	L+

This transaction will add a new *T*-account of "Income Tax Expense" and will change the balance of the "Accrued Expenses" *T*-account as follows:

Table 2-09(B) Example 2-07 Posting

Income Tax Expense		Accrued Expenses			
Ex 2-07	980	Ex 2-04-i	24,000	5,000	Ex 1-11
				24,000	Ex 1-12
				14,000	Ex 2-04-ii
				1,500	Ex 2-06
				980	Ex 2-07
				21,480	Balance

The following is the revised Adjusted Trial Balance that accommodates the tax and additional liability in the "Accrued Expenses" *T*-account:

Table 2-10 Revised Adjusted Trial Balance

Adjusted Trial Balance – B (with Income Tax)		
Accounts	**Debit**	**Credit**
Cash	438,000	
Accounts Receivable	60,000	
Inventory	34,800	
Prepaid Rent	85,000	
Prepaid Insurance	16,500	
FF&E	500,000	
Accumulated Depreciation		6,000

Table 2-10 (Continued)

Adjusted Trial Balance – B (with Income Tax)		
Accounts	**Debit**	**Credit**
Accounts Payable		55,000
Accrued Expenses		21,480
Unearned Revenues		–
Notes Payable		400,000
Common Stock – Rachel & Brad		450,000
Common Stock – Jane		200,000
Revenues		115,000
Cost of Sales	45,200	
Labor Expense	38,000	
Rent Expense	15,000	
Utility Expense	5,000	
Insurance Expense	1,500	
Depreciation Expense	6,000	
Income Tax Expense	980	
Interest Expense	1,500	
Total	1,247,480	1,247,480

2.5 PREPARATION OF FINANCIAL STATEMENTS

2.5.1 The Income Statement

If the taxable income is zero (0) or negative, income tax is not to be calculated. In other words, when a business does not have taxable profit, it does not pay tax. Once the tax is deducted, the final amount is the Net Profit of the business. The net profit amount of R&B Grill during the first month is $1,820. This should be transferred to the Balance Sheet into the Retained Earnings section of the Owners' Equity. The tentative Income Statement introduced at the end of Chapter 1 is obsolete because it showed only the temporary result with limited information. It was presented only to deliver the nature of accounting procedure at that point. After the operations of the first month, the performance of R&B Grill is presented in the official Income Statement as follows:

Table 2-11 The Income Statement After One Month

The Income Statement The R&B Grill For the month of January, 2xx1		
Revenues		115,000
Cost of Sales	45,200	
Labor Expense	38,000	
Rent Expense	15,000	
Utility Expense	5,000	
Insurance Expense	1,500	
Depreciation Expense	6,000	
Interest Expense	1,500	
Total Expenses		112,200
Income before Income Tax		2,800
Income Tax (35%)		980
Profit (Loss)		1,820

2.5.1.1 Analysis of the Operational Result with the Income Statement

The format of the Income Statement presented is simplified to deliver the important points of information. As we continue with additional details, more structured formats will be introduced in later chapters. The result presented on the Income Statement shows that the business has generated revenues of $115,000. Out of the revenues, the total expense incurred before taxes was $112,200, leaving $2,800 of taxable income. From this amount, after taking 35% ($980) for the Income Tax, the net profit was $1,820.

The list of expenses shows itemized accounts with the amount incurred. Paying attention to the details of these is critical to running a business successfully. The information can be further analyzed with percentage calculation by dividing every expense account by the revenues amount. Table 2.11 shows the percentage information of R&B Grill's monthly results. The management of R&B can compare the result with the following months to see whether the operation improves cost efficiency. It can also compare the result with other competitors this way.

Table 2-12 The Income Statement with Vertical Analysis

The Income Statement (with vertical analysis) The R&B Grill For the month of January, 2xx1			
Revenues		115,000	100.0%
Cost of Sales	45,200		39.3%
Labor Expense	38,000		33.0%
Rent Expense	15,000		13.0%
Utility Expense	5,000		4.3%
Insurance Expense	1,500		1.3%
Depreciation Expense	6,000		5.2%
Interest Expense	1,500		
Total Expenses		112,200	97.6%
Income before Income Tax		2,800	2.4%
Income Tax (35%)		980	
Profit (Loss)		1,820	1.6%

The Income Statement with the additional information presents the all expenses in percentage of the revenues. During the first month, the business spent 39.3% of its revenues for its inventory. This means that the product cost was 39.3 cents of every $1 of the revenues. By the same comparison, it spent 33 cents in labor for every $1 of revenues, and so forth. The last line that says "Total expenses" is the total amount of the expenses before Income Tax combined together and it was 97.6% of the revenues. As the result, the company has taken in 2.4% of its revenues as its taxable income (0.024 × $115,000 = $2,800). After paying 35% of it (0.35 × $2,800 = $980) for the Income Tax, the Net Profit is $1,820. This is the net profit R&B Grill earned in the first month. As introduced at the end of Chapter 1, this amount is added to the Owners' Equity under the title of "Retained Earnings." This means that the business has earned $1,820 during the month and added it to the owners' investments. This result will be presented in the Balance Sheet. This notion will be repeatedly mentioned as we move forward to advanced concepts and techniques.

2.5.2 The Balance Sheet

Once the Income Statement is prepared, the next statement to present is the Balance Sheet to show R&B Grill's financial position at the end of the first month:

Table 2-13 Balance Sheet after the First Month

The Balance Sheet The R&B Grill On January 31, 2xx1		
Assets		
	Cash	438,000
	Accounts receivable	60,000
	Inventory	34,800
	Prepaid rent	85,000
	Prepaid insurance	16,500
	FF&E	500,000
	Accumulated depreciation	(6,000)
Total Assets		**1,128,300**
Liabilities and Owners' Equity		
	Liabilities	
	Accounts payable	55,000
	Accrued expenses	21,480
	Unearned revenues	–
	Notes payable	400,000
	Total Liabilities	476,480
	Owners' Equity	
	Common stock – Rachel & Brad	450,000
	Common stock – Jane	200,000
	Retained Earnings	1,820
	Total Owners' Equity	651,820
Total Liabilities & Owners' Equity		**1,128,300**

Although the necessary information is presented in Table 2.13, the format looks different from the one presented in Chapter 1. There are two different formats to present the balance sheet. The current one is the reporting format, while the previous one is the account format. The account format divides the assets from the liabilities and equity horizontally. This format is easier to verify the accounting equation of $A = L + E$. The reporting format lists assets, liabilities, and equity vertically. The only difference is the shape of the information displayed. All other information is identical.

At the end of the first month, R&B Grill has $1,128,300 in its assets. Out of this amount, the company owes $476,480 to its creditors and its owners can claim $651,820 for their ownership. It must be recognized that the amount of owners' equity has increased by $1,820 from its beginning amount of $650,000. This increase is the contribution made by the operations with the net profit of $1,820.

On the first day, the business started with the asset amount of $800,000 (Table 2.2) with the liabilities of $150,000. Now its assets have grown by $328,300 to $1,128,300. However, the net growth of owners' equity from the operating result is only $1,820. Further investigation of the ending balance sheet shows that the company's liabilities grew by $326,480 to $476,480. Comparing the growth of total assets (which is $328,300) and that of the total liabilities (which is $.326,480), the difference is $1,820 (=total assets growth of $328,300 – total liabilities growth of $326,480). In other words, total assets have grown more than the liabilities by $1,820 that is the growth of the Owners' Equity. This is a brief example of how financial statements can explain the progress of the business. The next chapter will introduce more advanced methods of analyzing business progress.

CHAPTER 3

ANALYZING BUSINESS PROGRESS PRESENTED IN THE FINANCIAL STATEMENTS

3.1 INTRODUCTION

The 18 examples introduced in the first two chapters presented common but major activities of a restaurant business during its beginning stage. Along with the examples, the changes of the financial condition of the business were explained with a series of financial statements. In practice, however, businesses do not prepare their financial statements, particularly the Balance Sheet, as often as introduced in this book. This book presented it in such a manner only to help readers comprehend the big picture about how business activities transform to the financial information on both statements.

3.2 CHANGES OF FINANCIAL POSITION

This chapter provides further analyses of the progress of R&B Grill by explaining the changes of its financial positions over time. The following Table 3.1 displays the balance sheet records of R&B Grill in two columns before it starts its operations. The first column shows its status on its first day and the second column is on the day when it was prepared to start operating. These records are the same as those presented in Tables 1.2 and 1.12. The structure presented in the Table 3.1 followed the reporting format that shows Assets on the top followed by Liabilities and Owners' Equity. By presenting the Balance Sheet records this way, it is easy to view the progress of the business.

Table 3-01 The Changes of the Balance Sheets Records Over Time (During the Pre-opening))

The Balance Sheet R&B Caterer, Inc.	Day 1	Pre-opening
Assets:		
Cash	800000	415000
Accounts receivable		
Inventory		80,000
Prepaid rent		100,000
Prepaid insurance		18,000
FF&E		500,000
Total Assets	**800,000**	**1,113,000**
Liabilities & Equity		
Liabilities		
Accounts payable		55,000
Accrued expenses		
Unearned revenues		8,000
Notes payable	150,000	400,000
Total Liabilities	**150,000**	**463,000**
Owners' Equity		
Common stock		
– Rachel & Brad	450,000	450,000
Common stock		
– Jane	200,000	200,000
Total Owners' Equity	**650,000**	**650,000**
Total Liabilities & Equity	**800,000**	**1,113,000**

On its first day, R&B Grill, Inc. was created with the $800,000 of cash that was financed through one liability item (Notes Payable of $150,000) and the Owners' Equity of $650,000 invested by two owner groups – Rachel and Brad together as one group, and Jane. The company's total assets ($800,000) were the same as its total liabilities ($150,000) and total owners' equity ($650,000) combined. This is the financial position of R&B Grill on the first day.

The second column titled "Pre-opening" shows the result of the company's financial position after recording all transactions described in Examples 1.3 through 1.8. During this time, the company has not served even a single customer. Thus, there are no revenues or expenses involved. Through the preparation activities, R&B Grill has expanded the volume of its assets from $800,000 by $313,000 to $1,113,000. The liabilities section shows that the total amount also has grown from $150,000 to $463,000 by the same amount of $313,000. As shown, the growth of assets is completely financed by liabilities. Thus, its owners' equity has not changed at all. The changes of R&B Grill's assets were the result of the company's decisions of allocating its cash to obtain necessary material to prepare itself to open to serve customers. These can be viewed as the company's internal investing activities, which are also often called "resource allocation." To this point, the company had been preparing itself to start its operations.

3.3 THE IMPACT OF OPERATIONAL RESULTS ON FINANCIAL POSITION

The second analysis covers the company's financial position changes as the result of its operating activities. Examples 1.9 through 1.12 (in Chapter 1) and Examples 2.1 through 2.6 (in Chapter 2) are operating activities, and the result was presented in the Income Statement. Table 3.2 presents both the Balance Sheet records and the Income Statement records of the first month. The Balance Sheet records are presented in three columns that show the company's financial position on three different days. The first two columns are the same as what was presented in Table 3.1. They are presented again only to explain the changes of the firm's financial position during the first month.

Table 3-02 The Balance Sheets and the Income Statement of the First Month

	The Balance Sheet R&B Grill Day 1	Pre-opening	After 1 month	The Income Statement The R&B Grill For the month of January, 2xx1		
Assets:				Revenues		115,000
Cash	800,000	415,000	438,000	Cost of Sales	45,200	
Accounts Receivable			60,000	Labor Expense	38,000	
Inventory		80,000	34,800	Rent Expense	15,000	
Prepaid Rent		100,000	85,000	Utility Expense	5,000	
Prepaid Insurance		18,000	16,500	Insurance Expense	1,500	
				Depreciation Expense	6,000	
FF&E		500,000	500,000	Interest Expense	1,500	
Acc. Depreciation			(6,000)	Total Expenses		112,200
Total Assets	**800,000**	**1,113,000**	**1,128,300**	Income before Income Tax		2,800
				Income Tax (35%)		980
Liabilities & Equity				**Profit (Loss)**		**1,820**
Liabilities						
Accounts Payable		55,000	55,000			
Accrued Expenses		8,000	21,480			
Unearned Revenues			–			
Notes Payable	150,000	400,000	400,000			
Total Liabilities	**150,000**	**463,000**	**476,480**			

Owners' Equity			
Common Stock			
– Rachel & Brad	450,000	450,000	450,000
Common Stock			
– Jane	200,000	200,000	200,000
Retained Earnings			1,820
Total Owners' Equity	**650,000**	**650,000**	**651,820**
Total Liabilities & Equity	**800,000**	**1,113,000**	**1,128,300**

The changes of the financial position during the first month show that the entire assets started at $800,000 in the beginning that grew to $1,113,000 and finally, it became $1,128,300 at the end of the month. As pointed out above, the company has obtained many different asset items during the preparation. It must be noticed that one asset account (Accumulated Depreciation under the FF&E) in the balance sheet at the end of the month shows a negative amount. It is a Contra-Asset account that offsets the value of the FF&E over time by the amount presented. So, the net value of the FF&E at the end of January is $494,000 (although this net amount is not displayed). This deduction in value is caused by the use of the FF&E through the time.

During the same month, the firm's liabilities grew from $150,000 to $476,480. The details reveal that the Notes Payable is the major cause of this growth. In addition, the firm also carries two other liability items – one in the Accounts Payable and the other in the Accrued Expenses. Readers can refer to the Example 2.4 for detailed nature of the two accounts.

The owners' equity section shows no significant changes. However, the total amount of it grew by $1,820 at the end of the month with the introduction of a new account title of "Retained Earnings." As explained in the Examples 2.11 and 2.12, the result of the firm's operational result of each accounting period, presented as Net Profit (or Loss) on the Income Statement, must be returned to the investors. This is how the business increases owners' investments. The only way to increase the value of the owners' investments is through accumulating positive Retained Earnings, and the only way to achieve this goal is through achieving good operating results continuously over time. In the third part of this book where the concept and techniques of valuation is discussed, more details of this will be explained.

3.4 COMMONLY USED FORMAT OF FINANCIAL STATEMENTS IN THE INDUSTRY

The format of the financial statements introduced so far has been quite simple only with account titles introduced. Although simplicity helps to grasp the concept easy, it may be insufficient to use effectively for many different purposes. This chapter also introduces a standardized format of financial statements that is commonly used in the industry, presenting the same information. Later in this chapter, more advanced examples will be introduced to help readers to develop better knowledge to consolidate the business performance and its status.

3.4.1 Advanced Format of the Income Statement

The Income Statement introduced in the Table 3.2 shows a simplified list of accounts. For small operations, it may work fine with no difficulties. However, there is another way to present the same information for better analysis. Compare the format of the Income Statement introduced in the following Table 3.3 with the previous one presented in Table 3.2.

Table 3-03 The Income Statement in a Revised Format

The Income Statement The R&B Grill For the month of January, 2xx1	Amounts	V.A.%*
Revenues	115,000	100.0%
Cost of Sales	45,200	39.3%
Gross Profit	**69,800**	**60.7%**
Operating Expenses		
Labor Expense	38,000	33.0%
Rent Expense	15,000	13.0%
Utility Expense	5,000	4.3%
Insurance Expense	1,500	1.3%
Depreciation Expense	6,000	5.2%
Total Operating Expenses	**65,500**	**57.0%**
Operating Profits (EBIT)	**4,300**	**3.7%**
Interest Expense (Income)	1,500	1.3%
Income before Income Taxes (EBT)	2,800	2.4%
Income Tax (35%)*	980	
Net Income (loss)	**1,820**	1.6%

* V.A.% = Vertical Analysis %.

The report in the new format provides more specific and refined information. The Gross Profit line shows the amount left over after the Cost of Sales is deducted from the Revenues. The term Gross Profit indicates that the amount must be large enough to cover all other expenses that follow. The amount of all expenses spent in the operations, without the Cost of Sales, was $65,500. The vertical analysis column shows the percentage of individual expenses and profits of the revenues. This information will enable operators to enhance their cost control by taking appropriate actions on individual expenses by keeping track of them.

The restaurant industry has been updating their own reporting system, called "the Uniform Systems of Accounts for Restaurant." It is very similar to the format introduced above. The system is designed to maximize the usefulness of the financial information for the practitioners and other stakeholders. It still follows the generally accepted account principles (GAAP) for details. The Uniform Systems of Accounts for Restaurant (simply, "the uniform system" from now in this book) provides more specific information by separating revenues into subgroups such as "Food Sales" and "Beverage Sales." By doing so, it also separates the cost of sales into the "Cost of Food Sales" and the "Cost of Beverage Sales." This approach allows operators to separately track the individual lines of operations to obtain as accurate information as possible. Operators can modify this approach and develop more effective formats of their own. One example is dividing sales into "sales on premise" and "pick-up or delivery." Many fast food operators who deliver their products have adopted this format. Another example is to divide sales into different types of meal such as "breakfast sales," "lunch sales," and "dinner sales."

Dividing the sales by the line of operations is easy. However, allocating costs and expenses accordingly is hard. For this reason, the cost of sales is usually calculated comprehensively. The next important expense account in restaurant businesses is the "Labor Cost." These two expenses – Cost of Sales and the Labor Cost – are often called the "Prime Cost" in the industry. Labor Cost is also hard to be allocated accurately to specific lines of operations due to the overlapping effect of employees' task, but it can be divided, for example, into salaries and wages to identify the spending volume on each category. It also includes all other specific spending for employee-related activities, such as benefits, overtime payment, travel, and other similar accounts. All expenses can be divided into as many subcategories as possible for efficient control of the business. However, when they are presented in the final report, only the master account titles are presented like the example in Table 3.3. Specific individual titles of accounts are only used inside each business. They are not presented to the public.

Among the list of expenses displayed in the Income Statement in our examples, the group of expenses that are listed under the title of the "Operating expenses," from Labor Expenses to Depreciation Expense, led to the Operating Profits of $4,300. This is the amount of profit generated by the operating activities of R&B Grill during the first month. Many businesses also use the title of "Earnings before Interest and Taxes (EBIT)" for this as presented in the Table.

Beneath the Operating Profit (EBIT), non-operating expenses are listed. One of them is Interest and the other is Income Tax. Both are non-operating expenses. As pointed out, the interest expense is incurred due to the loan. Sometimes, a business may bring in interest income from the loans they have given to others. In such occasions, the receiving is an "income" rather than an "expense." Any expenses listed in the section of the expense in positive value are subtracted in calculation process. To add the value of "income," it must be presented in negative value. This is why the "Income" is presented in a parenthesis in Table 3.1.

Once the Interest Expense (Income) is subtracted from (or added to) the Operating Profits, the amount left over is used to calculate the Income Tax. For this nature, this title (Income before Income Taxes) is often called the "Taxable income." Many businesses present this amount under the title of the "Earnings before Taxes (EBT)." The Income tax amount is calculated by applying the tax rate to the Income before income taxes. In our example, the Income tax is calculated at 35% of the EBT.

3.4.2 Advanced Format of the Balance Sheet

The Balance Sheet can also be presented in a standardized format commonly used by the industry in a more efficient structure as shown in Table 3.4. Most companies in the restaurant industry present their Balance Sheets in the format presented in the Table 3.4.

First, the assets are divided into two groups as "Current Assets" and "Long-term Assets." The items included in the "Current Assets" are the resources that can be consumed within a year in operational activities. The business, by consuming them, will bring in cash. This nature is usually described in most accounting books as "the current assets are the items that can be converted into cash within one year." The "Account Receivable" can be collected; and the collected amount becomes "cash" of the business. The "Inventory" should be processed into the products for sales and "cash" will be brought in when sold. "Prepaid Expenses" do not usually bring in cash. Because they are deposits made for future expenses, however, they will save cash when they are used for expenses. This is how current asset items are "converted" into cash within 1 year. The following is the revised format of the Balance Sheet of R&B, followed by a brief analysis of its financial position.

Table 3-04 The Balance Sheet in a Revised Format

The Balance Sheet R&B Caterer, Inc.	Pre-opening	After 1 month
Assets:		
Current Assets		
Cash	415,000	438,000
Accounts receivable		60,000
Inventory	80,000	34,800
Prepaid Rent	100,000	85,000
Prepaid insurance	18,000	16,500
Total Current Assets	**613,000**	**634,300**
Long-term Assets		
FF&E	500,000	150,000
Less Acc. Depreciation		(2,000)
Leasehold Improvement		250,000
Less Acc. Depreciation		(2,000)
Smallware		100,000
Less Acc. Depreciation		(2,000)
Net FF&E	**500,000**	**494,000**
Total Long-term Assets	**500,000**	**494,000**
Total Assets	**1,113,000**	**1,128,300**

Table 3-04 (Continued)

	Pre-opening	After 1 month
Liabilities & Equity		
Liabilities		
Current Liabilities		
Accounts payable	55,000	55,000
Accrued expenses		21,480
Unearned revenues	8,000	--
Total Current Liabilities	**63,000**	**76,480**
Long-term Liabilities		
Notes payable-A	450,000	150,000
Notes payable-B		250,000
Total Long-term Liabilities	**450,000**	**400,000**
Total Liabilities	**513,000**	**476,480**
Owners' Equity		
Common stock		
– Rachel & Brad	450,000	450,000
Common stock		
– Jane	200,000	200,000
Retained Earnings		1,820
Total Owners' Equity	**650,000**	**651,820**
Total Liabilities & Equity	**1,163,000**	**1,128,300**

3.4.3 Explanation of the Revised Balance Sheet Information After the First Month

After 1 month, R&B Grill still has $634,000 in the Current Assets. This should be compared with the Current Liabilities. The Current Liabilities are the amount of debt (or obligation) that must be paid within 1 year. As mentioned earlier, however, these obligations must be handled within the next couple of weeks in most cases. This company has $76,480 in its current liabilities, while it carries $634,000 in its current assets. The cash amount available in the business is $438,800. The capability of a firm to pay its current liabilities is called the "Liquidity." R&B Grill, according to the given information, has enough cash on hand for its immediate debt, which indicates it has strong liquidity. It must be explained for the excessive amount of cash on hand. As R&B Grill just started its business as a new company, it is highly recommended to preserve enough cash for future contingency. Restaurant business has been known for its high risk of failure, and many experts have pointed out the lack of cash on hand as the primary reason. For potential financial difficulties that the company may face in the future, Rachel and Brad decided to keep enough cash in a separate account. This is why the current balance of cash is excessively high. Many restaurant firms usually keep this type of emergency funds in a separate account called "Short-term Investments" or "Marketable Securities." R&B Grill is recommended to invest its extra cash into short-term governmental bonds or other similar financial securities that will earn interest income in the future.

Under the Current Assets are the Long-term Assets that refer to the resources that can be used by the business for longer than a year. These include, but are not limited to, buildings, land, furniture and/or equipment (FF&E), and other similar items. Many practitioners call these as "Fixed assets." Except land, all Long-term Assets depreciate over time. It has been introduced how to estimate the amount of depreciation expenses using the straight-line method. As mentioned earlier, there are other methods. One of them is called the "usage method" that calculates the depreciation by counting the actual output of the item. Vehicles are a good example of this method. Every mile driven will deduct from the value of the vehicle at the predetermined rate such as 45 cents per mile, for example. Using this method, the more a vehicle is driven, the more depreciation expense will be incurred. The other method is called the "accelerated method." This method applies a larger amount of depreciation expense during the early years, and the amount of depreciation gradually decreases as the time passes by until the item reaches its predetermined residual value. Technical details of these methods are not introduced in this book because they are not conceptually significant. Readers may refer to other sources (books or online) to find out necessary technical details.

What is important about depreciation expense is that it is a "Non-cash expense," which means that, different from other expenses, it does not involve actual cash payments. Because there is no cash payment involved with depreciation expenses, the cash amount presented for it remains in the business. However, as an expense, it reduces the amount of the taxable income (EBT).

In summary, depreciation expenses are incurred on the long-term assets obtained as the result of investing activities of the business itself. Once long-term assets are obtained, the business presents the estimated amount of depreciation expenses. However, as they are non-cash expenses, the business does not make the payment of it. Instead it keeps the cash. Due to this nature, depreciation expenses over time will be accumulated in the business. By doing so, the business can recover its investments in time. Regardless the different methods of calculating this expense, when the amount of depreciation is added to the net profit to determine the cash earned through operations, the total amount (net profit plus depreciation) becomes the same. More details of this effect will be explained when the Cash Flows are introduced in Chapter 7.

The Balance Sheet presented in Table 3.4 shows a restructured itemized list of the FF&E of R&B Grill. It is revised to accommodate more details to present more accurate information of the R&B Grill's financial position. The original FF&E *T*-account of $500,000 is now divided into three separate long-term asset accounts – FF&E, Leasehold Improvement, and Smallware. FF&E stands for Furniture, Fixture, and Equipment. Leasehold Improvement is the amount spent to convert the leased space into a restaurant by installing power cables, pluming, carpet and flooring, signage, and many other supporting devices. Smallware represents such items as plates, cups and glasses, silverware, linen, and other similar items. It must be noticed that each account is followed by individual Accumulated Depreciation *T*-accounts. This is explained further in the sixth problem in the Example 3.1.

The "Long-term Liabilities" are listed under the Current Liabilities. Long-term Liabilities include the liability items that can be paid over 1-year time. Notes Payable account is a good example. Usually these items incur interest expenses that are presented in the Income Statement. These debts, however, usually require a long series of monthly payments of a fixed amount that includes the principal and the interest. This is called the "debt service." The technique of determining the fixed amount of debt service will be introduced in the third part of this book. In the revised Balance Sheet (Table 3.4), one comprehensive Notes Payable account is divided into two separate accounts of N/P-A and N/P-B. This will be introduced with more details in the eighth problem in the Example 3.1.

Table 3-05 Beginning Balances of T-accounts in the Year 1

Asset accounts	Cash	Beg. Bal. 438,000	
Liability accounts	Accounts payable		55,000 Beg. Bal
	Accrued expenses		20,500 Beg. Bal
Equity accounts	Common stock – Rachel & Brad		450,000 Beg. Bal
	Common stock – Jane		200,000 Beg. Bal
Revenues & Expenses	Revenues		
	Cost of sales		

Accounts receivable

Debit	Credit
Beg. Bal. 60,000	

Unearned revenues

Debit	Credit
	– Beg. Bal

Labor expenses

Debit	Credit

Inventory

Debit	Credit
Beg. Bal 34,800	

Notes payable – A

Debit	Credit
	150,000 Beg. Bal

Restaurant operating expenses

Debit	Credit

Table 3-05 (Continued)

Prepaid rent

Beg. Bal	85,000

Notes payable – B

	250,000 Beg. Bal

Utility expenses

Prepaid Insurance

Beg. bal 16,500	

Insurance expense

FF&E

Beg. Bal	500,000		

Accumulated Depreciation – FF&E

		6000	Beg.Bal

Administrative expenses

Repair and maintenance expenses

Table 3-05 (Continued)

Rent expenses	
Interest expenses	

The revised formats of R&B Grill's financial statements can provide readers with specific information that can analyze the performance and the status of the business using the subtotals on each segment. It is time to move to more advanced applications. Let us practice with the following examples of the next year's business activities. Example 3.1 is provided to help readers review and practice necessary accounting concepts and techniques introduced so far. It is strongly recommended that readers follow through the given problems thoroughly and prepare their own financial statements. For practice purpose, miscellaneous calculations are eliminated. Readers are encouraged to post these records to each *T*-account. The solution table (of the *T*-accounts) will be provided later for verification. Make sure that the beginning balances are recorded correctly.

3.5 SUMMARY OF ANNUAL OPERATIONS

Example 3.1: *R&B Grill, Inc. 1 Year Later*

Rachel and Brad continued their business. After the first month of operations, they made an official start of their business by reorganizing the structure of record-keeping process with more breakdowns. One year after, the summary of their business records are as follows.

1. Their business opened 365 days. The number of total guests served turned out to be 53,650, and the total revenues of the year were $1,444,195. Their records show that, out of the total sales, $1,240,000 was received in cash and the rest of it was put on account at the time of sales. The accounting information in this example contains the revenues, cash, and accounts receivable. The operational information of the number of days and customer counts will be used later in forecasting and budgeting activities. The journal entry follows the same rules that have been used before.

Example 3-01-1 Journal Entry

Debit			Credit		
Impact	Account	Amount	Account	Amount	Impact
A+	Cash	1,240,000	Revenues	1,444,195	Rev/E+
A+	Accounts Receivable	204,195			

2. Further investigations revealed that out of the entire amounts of the Accounts Receivable, $164,195 has been collected. This is a simple adjustment of the collection of outstanding receivable amounts.

Example 3-01-2 Journal Entry

Debit			Credit		
Impact	**Account**	**Amount**	**Account**	**Amount**	**Impact**
A+	Cash	164,195	Accounts Receivable	164,195	A–

3. The total amount of inventory purchased during the year was $482,000. Of the entire inventory purchases, $54,000 has not been paid yet. All the rest has been paid in cash. This is a simple transaction of inventory purchase with partial payment. The amount of the cash payment must be calculated.

Example 3-01-3 Journal Entry

Debit			Credit		
Impact	**Account**	**Amount**	**Account**	**Amount**	**Impact**
A+	Inventory	482,000	Cash	428,000	A–
			Accounts Payable	54,000	L+

4. The physical inventory on hand at the end of the year was calculated to be $32,500. This transaction is about calculating the Cost of Sales. The amount of purchase is provided in the previous problem (#3).

Beg. Inventory + Purchases – Ending inventory = Cost of sales

$34,800 + $482,000 – $32,500 = $484,300

Example 3-01-4 Journal Entry

Debit			Credit		
Impact	**Account**	**Amount**	**Account**	**Amount**	**Impact**
Exp	Cost of Sales	484,300	Inventory	484,300	A–

5. The following is the summary of other expenses.

This problem contains several new expense accounts. They must be created. Each expense item is journalized in the debit side as usual. However, the credit entries require calculating the amount and the title of account for the payment. Most of them were paid in cash. So, the total amount of cash can be journalized in a lump sum at the end. It must be noticed that some expenses used deposited funds (i.e., prepaid expenses), while some were put on in the account.

First, put all the expenses separately on the debit side and then, address individual deposited funds used. Then, record the amount put in the account. After that, calculate the amount of cash paid.

a: Total payroll expenses: $401,800

b: Total occupancy expenses: $ 63,295

c: Total marketing expenses: $ 16,770

d: Total utility expenses: $ 33,750

e: Total insurance expense $ 18,000

(*This includes Prepaid Insurance. The balance of the account is completely used up.)

f: Total administrative expenses: $ 79,555

g: Total repair & maintenance: $ 14,820

h: Total Rent expenses: $180,000

(*This includes Prepaid Rent. The balance of account is used up.)

i: Out of the entire expenses listed above, $72,000 is accrued to be paid within the first two weeks of the next year. The unpaid amount uses "Accrued Expenses."

(*Additional guide to calculate the amount of cash spent.)

The total expense amount is $825,900. Of that amount, subtract the beginning balances of Prepaid Expenses – Prepaid Insurance and Prepaid Rent. Then, subtract the amount of Accrued Expenses (that is put on account). The rest is $825,900. This is the amount cash payment.

Example 3-01-5 Journal Entry

trx #	Debit Impact	Account	Amount	Credit Account	Amount	Impact
a	Exp	Payroll expenses	401,800	Prepaid ins.	16,500	A–
b	Exp	Occupancy expenses	63,295	Prepaid rent	85,000	A–
c	Exp	Marketing exp.	16,770	Accrued expenses	72,000	L+
d	Exp	Utility expenses	33,750	Cash	634,490	A–
e	Exp	Insurance exp.	18,000			
f	Exp	Administrative exp.	79,555			
g	Exp	Repair & maint.	14,820			
h	Exp	Rent expenses	180,000			

The list of expenses introduced in this problem followed the group titles commonly used in the industry. Individual titles represent multiple accounts that belong to them. For example, the Labor Expense account includes salaries, wages, and employee benefits such as insurance. The list is also designed for continuous use in later chapters for further analysis of costs.

6. The amount of depreciation expenses is divided into cover-specific items as listed in following table. This breakdown was briefly introduced earlier with the revised format of the Balance Sheet (Table 3.4).

Account	Purchase Cost	Useful life	Salvage Value	Annual Depreciation
a: FF&E	$150,000	5 years	$30,000	$24,000 ($2,000/month)
b: Leasehold Improve	$250,000	10 years	$10,000	$24,000 ($2,000/month)
c: Smallware	$100,000	4 years	$4,000	$24,000 ($2,000/month)
Total	**$500,000**			**$72,000 ($6,000/month)**

This example shows the individual items of long-term assets that were previously recorded in the FF&E. The previous table is presented to provide the readers more specific details of group account titles. The total purchase cost was recorded in the previous Balance Sheets (Table 3.4). The new Balance Sheet will present itemized list with individual Accumulated Depreciation. It must be noticed that the total amount of individual depreciation expenses remains the same.

The table also provided the annual depreciation based on the salvage value (or residual value). Leasehold improvement is the cost of the contract work put into the property. The spending for it is considered as an investment made by the company. For this reason, it is recorded as a long-term asset. The process of putting the spending to the Balance Sheet is called "capitalization." R&B Grill has "capitalized" its spending of $250,000 on contract work.

Example 3-01-6 Journal Entry

	Debit			Credit		
trx #	Impact	Account	Amount	Account	Amount	Impact
a	Exp+	Depreciation – FF&E	24,000	Accu. Deprec. – FF&E	24,000	A–
b	Exp+	Depreciation – leasehold	24,000	Accu. Deprec. – leasehold	24,000	A–
c	Exp+	Depreciation – smallware	24,000	Accu. Deprec. – smallware	24,000	A–

* Please pay attention to the changes of the individual Accumulated Depreciation *T*-accounts.

7. The Accrued Expenses of $21,480 (on the beginning Balance Sheet) were paid on time in cash. This is a simple adjusting journal entry of payment of the accrued expense. The impact of this can be easily seen when the *T*-account of the Accrued Expenses is posted.

Example 3-01-7 Journal Entry

Debit			Credit		
Impact	Account	Amount	Account	Amount	Impact
L-	Accrued Expenses	21,480	Cash	21,480	A–

8. The outstanding Account Receivable of $60,000 (in the beginning Balance Sheet) was all collected in cash. This is another simple adjusting entry of collecting the outstanding receivable. The impact on the Accounts Receivable can be easily noticed when the posting is completed.

Example 3-01-8 Journal Entry

Debit			Credit		
Impact	**Account**	**Amount**	**Account**	**Amount**	**Impact**
A+	Cash	60,000	Accounts Receivable	60,000	A–

9. Although the contract says the payment of the debt and interest could be paid from the second year, Rachel and Brad decided to start the payments from the first year to be debt-free as early as possible and made monthly payments. The following is the records of total payments (total interest and the principal) each long-term debt during the first year. It also shows the current balance after the payments of the first year.

Accounts	Loan	TTL payment	Interest	Principal	Balance
N/P – A	$150,000	$ 38,000	$18,000	$ 20,000	$130,000
N/P – B	$250,000	$140,000	– 0 –	$140,000	$110,000

This transaction divides the consolidated Notes Payable record (introduced in Chapter 1) into two separate accounts. Each long-term debt (Notes Payable) is separately presented in the Accounts column. "Loan" is the total amount borrowed, and "TTL payment" is the amount of the principal and interests paid on each debt per year. "Interest" is the expense paid for the loan. "Principal" is the net amount of debt paid after the interest. For the N/P-A, the actual amount of the debt payment was only $20,000 because $18,000 was used to pay the interest incurred. For N/P-B, as mentioned in the original purchase contract of the FF&E (for the N/P-B), this long-term debt does not incur interest expense. Thus, the entire amount of the payment ("TTL payment") of the N/P-B reduced the debt. Finally, "Balance" shows the remaining amount of each debt.

The journal entry of this transaction must include the cash payment, interest, and the actual payment of the debt as displayed in Table 3.1(9).

Example 3-01-9 Journal Entry

	Debit			Credit		
trx #	**Impact**	**Account**	**Amount**	**Account**	**Amount**	**Impact**
a	Exp	Interest Expense	18,000	Cash	178,000	A–
b	L-	Notes Payable-A	20,000			
c	L-	Notes Payable-B	140,000			

The impact on each Notes Payable accounts will be easily identified when the posting is completed.

Although all business-related transactions have been introduced, there is still one more transaction that must be included, which is the Income Tax. As decided previously, 35% of the taxable income (EBT) will be applied to the Income Tax of the first year. If readers calculate the taxable income of R&B Grill in the first year, it will be $21,667. This amount can be verified once the following transaction is completed and the Income Statement is prepared.

10. The Income Tax of the first year has been calculated to be $21,667. This will be paid when the tax report is prepared. Until then, the unpaid amount of the tax is included into the "Accrued Expenses" account. The journal entry of this is displayed in Example 3.10.

Example 3-01-10 Journal Entry

Debit			Credit		
Impact	**Account**	**Amount**	**Account**	**Amount**	**Impact**
Exp	Income Tax Expense	21,677	Accrued Expenses	21,677	L+

When all the journal entries in this example are correctly posted, the *T*-accounts will look like those in the following Example 3.1(10) "Solution of T-accounts."

Example 3-01-11 Solution of T-accounts – After the First Year

Asset accounts				Liabilities & Equity accounts				Revenues & Expenses			
Cash				**Accounts Payable**				**Revenues**			
beg. Bal.	438,000	428,000	3-01-3			55,000	beg. Bal			1,444,195	3-01-1
3-01-1	1,240,000	634,490	3-01-5			54,000	3-01-3				
3-01-2	164,195	21,480	3-01-7			109,000	bal.				
3-01-8	60,000	178,000	3-01-9								
sub total	1,902,195	1,261,970	sub total					Cost of Sales			
bal.	640,225							3-01-4	484,300		
				Accrued Expenses							
Accounts Receivable				3-01-7	21,480	21,480	beg. Bal				
beg. Bal.	60,000	164,195	3-01-2			72,000	3-01-5i	Labor Expenses			
3-01-1	204,195	60,000	3-01-8			21,667	3-01-10	3-01-5a	401,800		
sub total	264,195	224,195	sub total	sub total	21,480	115,147	sub total				
bal.	40,000					93,667	bal.				
								Occupancy Expenses			
								3-01-5b	63,295		

Inventory

beg. Bal	34,800	484,300	3-01-4
3-03-3	482,000		
sub total	516,800	484,300	sub total
bal.	32,500		

Notes Payable – A

3-01-9b	20,000	150,000	beg. Bal
		130,000	bal.

Marketing Expenses

3-01-5c	16,770

Prepaid Rent

beg. Bal	85,000	85,000	3-01-5h
bal.	–		

Notes Payable – B

3-01-9c	140,000	250,000	beg. Bal
		110,000	bal.

Utility Expenses

3-01-5d	33,750

Prepaid Insurance

beg.bal	16,500	16,500	3-01-5e
bal.	–		

Insurance Expenses

3-01-5e	18,000

Example 3-01-11 (Continued)

FF&E

beg. Bal	150,000		

Accumulated Depreciation – FF&E

		2,000	beg. Bal
		24,000	3-01-6a
		26,000	bal.

Leasehold Improvement

beg. Bal	250,000		

Common Stock – Rachel & Brad

		450,000	beg. Bal

Common stock – Jane

		200,000	beg. Bal

Administrative Expenses

3-01-5f	79,555		

Repair & Maintenance Expenses

3-01-5g	14,820		

Rent Expenses

3-01-5h	180,000		

Depreciation Expense – FF&E

3-01-1a	24,000		

Accumulated Depreciation – Leasehold

		2,000	beg. Bal
		24,000	3-01-6b
		26,000	bal.

Smallware

beg. Bal	100,000

Accumulated Depreciation – Smallware

		2,000	beg. Bal
		24,000	3-01-6c
		26,000	bal.

Depreciation Expense – Leasehold improve.

3-01-1b	24,000

Depreciation Expense – Smallware

3-01-1c	24,000

Interest Expenses

3-01-9a	18,000

Now that all *T*-accounts have been updated with their balances, let's organize them directly into the Income Statement and the Balance Sheet. When the balances of the revenue and expense are presented, the Income Statement will be as follows:

Table 3-06 The Income Statement in a Revised Format

The Income Statement The R&B Grill	1st month	VA%*	Next 1 year	VA%*
Revenues	115,000	100.0%	1,444,195	100.0%
Cost of Sales	45,200	39.3%	484,300	33.5%
Gross Profit	69,800	60.7%	959,895	66.5%
Operating Expenses				
Labor Expenses	38,000	33.0%	401,800	27.8%
Restaurant Operating Expense		0.0%	63,295	4.4%
Marketing Expenses		0.0%	16,770	1.2%
Energy (Utility) Expenses	5,000	4.3%	33,750	2.3%
Insurance Expenses	1,500	1.3%	18,000	1.2%
Administrative Expenses		0.0%	79,555	5.5%
Repair & Maintenance Expenses		0.0%	14,820	1.0%
Rent Expenses	15,000	13.0%	180,000	12.5%
Depreciation – FF&E	2,000	1.7%	24,000	1.7%
Depreciation – Leasehold Improvement	2,000	1.7%	24,000	1.7%
Depreciation – Smallware	2,000	1.7%	24,000	1.7%
Total Operating Expenses	65,500	57.0%	879,990	60.9%
Operating Profits	4,300	3.7%	79,905	5.5%
Interest Expense (Income)	1,500	1.3%	18,000	1.2%
Income Before Income Taxes	2,800	2.4%	61,905	4.3%
Income Taxes (35%)	980		21,667	
Net Income (Loss)	1,820	1.6%	40,238	2.8%

* VA% is the percentage of individual expenses of the revenues of each year.

Once the Income Statement is prepared to show the final result of the operations, as explained earlier, the Income Tax amount of $21,667 can be verified and the net profit is $40,238. When this amount is transferred to the Balance Sheet, the Retained Earnings increase to $42058.

In accounting practice, another statement is prepared for this purpose. The title is Statement of Owners' Equity for personal businesses. Corporations use more specific title of "The Statement of Retained Earnings." Regardless of the title, the function of the report is to update the Owners' Equity section of the Balance Sheet and one of the important functions is to update the Retained Earnings by adding the operational result of the current fiscal accounting period.

Table 3-07 The Balance Sheet After One Year

The Balance Sheet R&B Grill	After 1 month	After 1 year
Assets:		
Current Assets		
Cash	438,000	640,225
Accounts Receivable	60,000	40,000
Inventory	34,800	32,500
Prepaid Rent	85,000	–
Prepaid Insurance	16,500	–
Total Current Assets	634,300	712,725
Long-term Assets		
FF&E	150,000	150,000
Less Acc. Depreciation	(2,000)	(26,000)
Leasehold Improvement	250,000	250,000
Less Acc. Depreciation	(2,000)	(26,000)
Smallware	100,000	100,000
less Acc. Depreciation	(2,000)	(26,000)
Net FF&E	494,000	422,000
Total Long-term Assets	494,000	422,000
Total Assets	1,128,300	1,134,725
Liabilities & Equity		
Liabilities		
Current Liabilities		
Accounts Payable	55,000	109,000
Accrued Expenses	21,480	93,667

Total Current Liabilities	76,480	202,667
Long-term Liabilities		
Notes Payable-A	150,000	130,000
Notes Payable-B	250,000	110,000
Total Long-term Liabilities	400,000	240,000
Total Liabilities	476,480	442,667
Owners' Equity		
Common Stock		
– Rachel & Brad	450,000	450,000
Common Stock		
– Jane	200,000	200,000
Retained Earnings	1,820	42,058
Total Owners' Equity	651,820	692,058
Total Liabilities & Equity	1,128,300	1,134,725

* The majority of this cash is reserved for emergency.

3.6 ANALYSIS OF R&B GRILL'S PERFORMANCE

Now that a series of business activities have been introduced from the accounting perspective up to this point, it is necessary to complete this chapter with a brief analysis about how R&B Grill has performed. The following is a brief analysis of the overall performance of the R&B Grill during its first two accounting periods – first month and 1 year after that.

The final Income Statement in the Table 3.6 shows two separate operational results: one for the first month and the second for the next year. It is not appropriate to compare the business results of 1 month with 1 year directly with the dollar amounts of revenues and expenses. However, the operational efficiency can be compared by converting the expenses into the percentage of revenues. The operational efficiency is about how tightly the management has controlled its expenses to generate its revenues. By comparing the percentage of each expense in different periods, it can be measured whether the business is improving its efficiency or not. Hence, businesses must keep using the same account titles continuously.

During the first month, R&B Grill has achieved and returned 1.6% of the revenues to the owners. During the next year, it has earned returned 2.8% to the owners. With this information only, the overall efficiency of the business operations has apparently improved. Further analysis reveals that its cost control, which is reflected in the Cost of Sales percentage, has declined from the 39.3% during the first month to 33.5% in the next year. This indicates that the business has improved its inventory control and it has increased the gross Profit by almost 6%. Basic analysis can be conducted this way by comparing the changes of each expense in its percentage over the revenues. This approach is helpful to determine whether the business is improving its operational efficiency over time. However, there are more advanced techniques. The next chapter will introduce them using the same sets of information.

PART II

HOW TO USE FINANCIAL INFORMATION FOR FORECASTING AND PLANNING

CHAPTER 4

RATIO ANALYSIS: ADVANCED TOOLS TO ANALYZE BUSINESS PERFORMANCE

4.1 INTRODUCTION

The previous three chapters have explained how business activities are transformed into meaningful information through the accounting process. At each step of the process, it has been illustrated how to view and analyze the changes of business conditions reflected on relevant financial statements. In this context, the purpose of accounting, which goes far beyond just the recording process, is clearly described in the definition of accounting presented by the American Accounting Association as "the system of identifying, measuring, and communicating economic information to permit informed judgments and decisions by users of the information." Accounting information must be communicated to the user groups to make sound decisions. This chapter, and the following ones, will cover the communicating function of accounting information so that the managers, as the first-hand user group, can make informed judgments.

4.2 FLOWS OF WEALTH TO AND FROM A BUSINESS

Because a business can be viewed as a set of convergent activities conducted to achieve the goal of adding value to the original investments made by the owners, the management's goal must be defined as making the business more valuable over time. In doing this, providing good profits is important but it is only a small fraction of the whole picture. The entire business activities can be presented in the form of a flowchart that illustrates the flow of wealth starting from the investors through the operating activities and finally back to the investors. Figure 4.1 "The Dynamics of Business Reflected on its Financial Statements" shows the structure of the flow of wealth presented in financial statements. It starts from the Balance Sheet on the top left. The assets of a business, which have been contributed by the creditors (liabilities) and owners (equity), are constantly used in the operating activities. The way a business conducts its operating activities is described in Figure 4.1 as its operational systems. The primary output of this operational system is the revenues. However, as we have noticed, expenses must be taken out of the revenues. So, the result of the operational system is the profit, which is described as the secondary output of the operational system. Only then, the final result, which is the secondary output, of the operational system

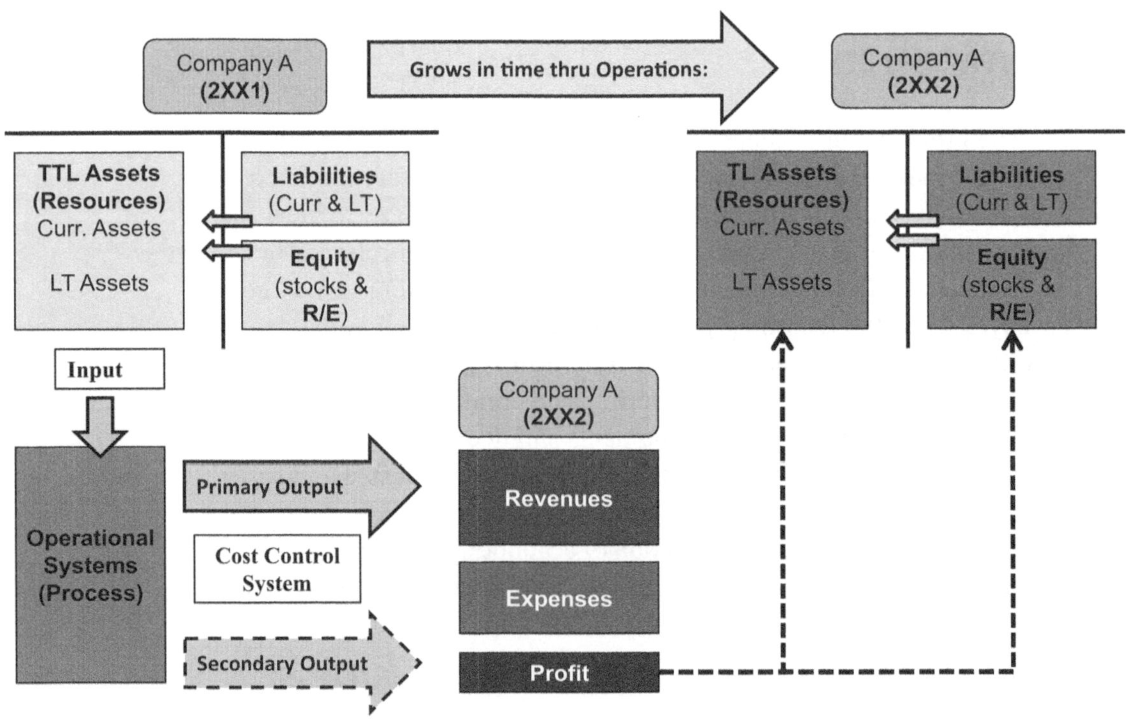

Figure 4.1 Wealth flows of a business – general.

is added to the owners' equity in the second balance presented in the top right corner. It must be emphasized here that the assets and liabilities have already been affected by the revenues and expenses. Put differently, operating activities that generate revenues and incur expenses have impacts on assets and liabilities. This has been explained in the previous chapters with impacts. When revenues are journalized, either the company's assets increase or its liabilities decrease. Whenever expenses are journalized, either the liabilities of the business increase or its assets decrease. The equity, however, is never affected directly by the revenues or expenses. Only the profits (or losses) as the result of the operations affect the owners' equity directly.

The generic structure of the flow of wealth illustrated in the Figure 4.1 can be redesigned with more detailed information as follows.

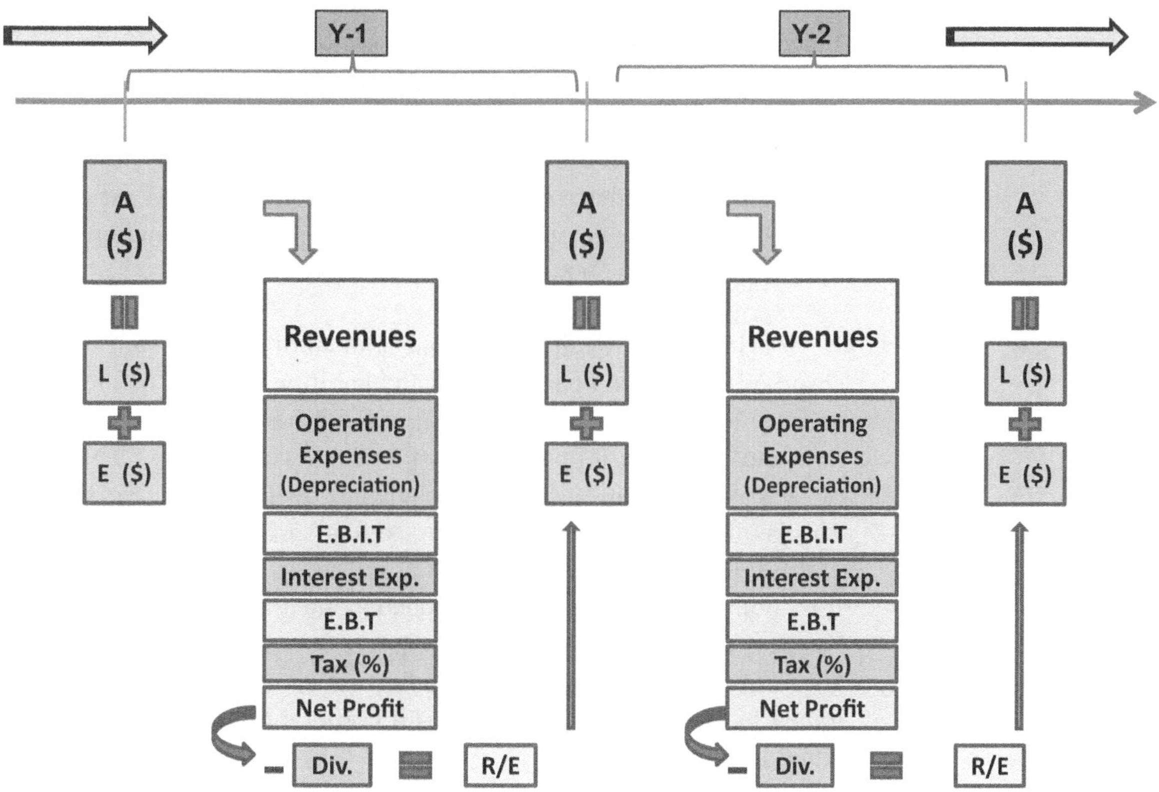

Figure 4.2 Wealth flows of a business with details.

Figure 4.2 is an advanced display of Figure 4.1 with more structured information of the Income Statements over multiple accounting periods. It represents the progress of a business for 2 years. The period of "Y-1" indicates the first accounting in a year. Naturally, the second term "Y-2" stands for the second year of the business.

The Balance Sheets are presented with the same components of assets, liabilities, and equity. The Income Statements show revenues and specific expense items such as Operating Expenses, Interest Expenses, and Taxes. After a certain group of expenses, different profit titles are presented. For example, after taking out the Operating Expenses, the amount left over is called "EBIT," which is an acronym of "Earnings before Interest and Taxes." This is often presented as the "Operating Profits." This is the amount of profit the business has obtained after incurring all expenses directly involved in its operations. The amount of this profit still includes the additional of expenses of "Interests" and "(Income) Taxes."

Naturally, after listing the amount of Interest Expenses, the remaining amount is the profit that still includes tax expenses. This is why it is called the "EBT," which stands for the "Earnings before (income) Taxes." The EBT is also called the Taxable Income. A business pays its income tax at a predetermined rate on the amount of its EBT. After taking out the tax expense amount from the EBT, the business can finally show the Net Profit on the bottom line of its Income Statement.

It is important to reiterate that the amount of the Net Profit is already absorbed into the company's assets and liabilities through recording revenues and expenses. The Net Profit, as the difference between the revenues and expenses, is finally returned to the Owners' Equity. This is how owners' investments increase through the operations of the business. This is often described as the "Value-adding" process of business. In this process, the owners may receive their returns in cash from the business. It is another way of returning the added value to the investors. In private businesses, owners may receive compensations in cash. In the corporate form of business, this return is called the dividends. Dividends are determined by the Board of Directors of a corporation and can be provided either in cash or stocks.

Then, the rest of the profits, after paying out cash returns or dividends, are put back (or reinvested, in other words) to the company under the "Retained Earnings." It is used for future business, which is the concept of reinvestment. Let's put our example company – R&B Grill – into this format to explain with numerical information. With all the results obtained in Chapter 3, R&B Grill's situation can be illustrated as follows:

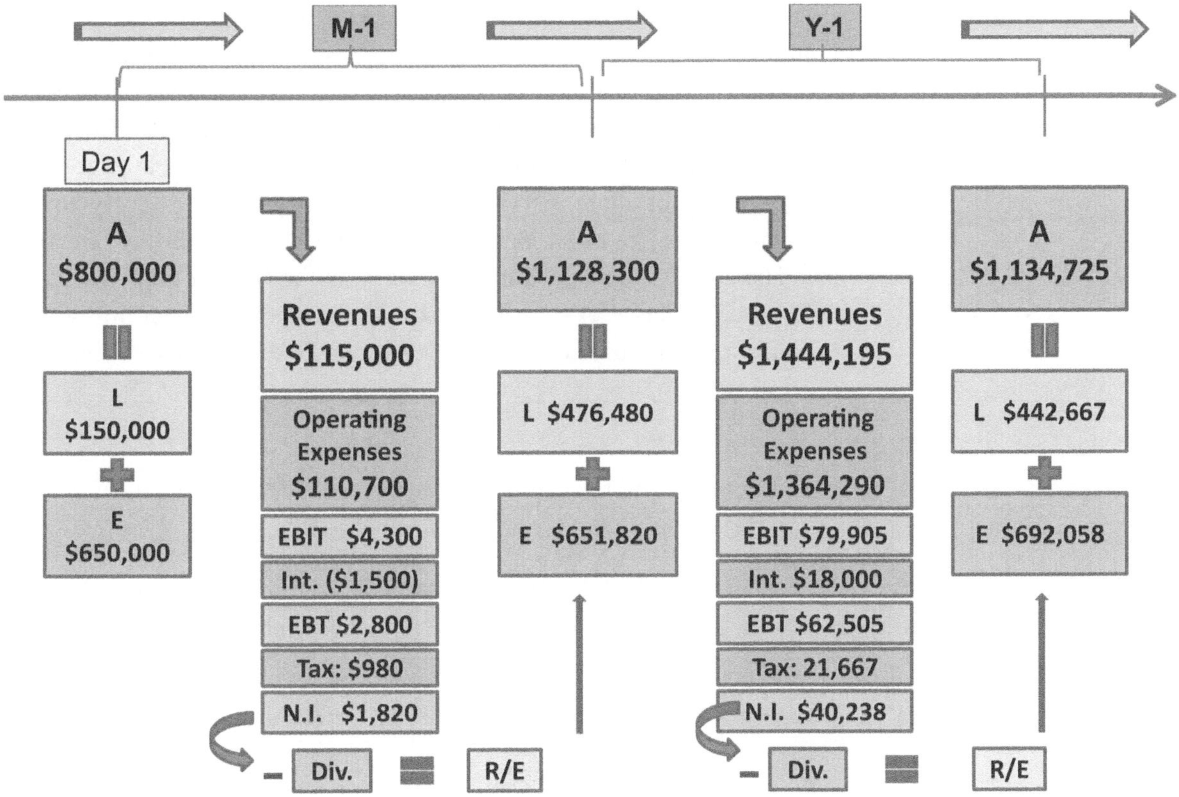

Figure 4.3 Wealth flows of a business – R&B Grill.

Figure 4.3 presents the periodical results of all the business activities introduced and explained so far in the sequence of financial statements. From the left side, the beginning balance sheet records of the first day are presented followed by the result of its operational activities of the first month. The contribution of the first month's operations is reflected in the ending balance sheet, which becomes the beginning balance sheet of the next term (which is the Y-1 column in the Figure). When financial statements are presented in this way, it becomes much easier to identify the trends of the business at a glance. The growing amounts of assets, equity, and revenues and profits along with the decreasing amount of liabilities clearly indicate that R&B Grill is moving in a positive direction.

4.3 VERTICAL AND HORIZONTAL ANALYSIS

There are two basic approaches to view the progress: one is to view the trends of each component such as revenues, expenses, or profits over time; and the other is to view the efficiency of each activity by comparing expenses and profit over the revenues. The former is called the horizontal analysis and the latter is called the vertical analysis. Both views focus on the changes of the same item from two different perspectives. By analyzing them this way, we can identify the initial growth patterns of the business. The analyses can be conducted both on the Income Statements and the Balance Sheets. Let's try this first with the Balance Sheet information of R&B Grill. The balance sheet records can be displayed as follows:

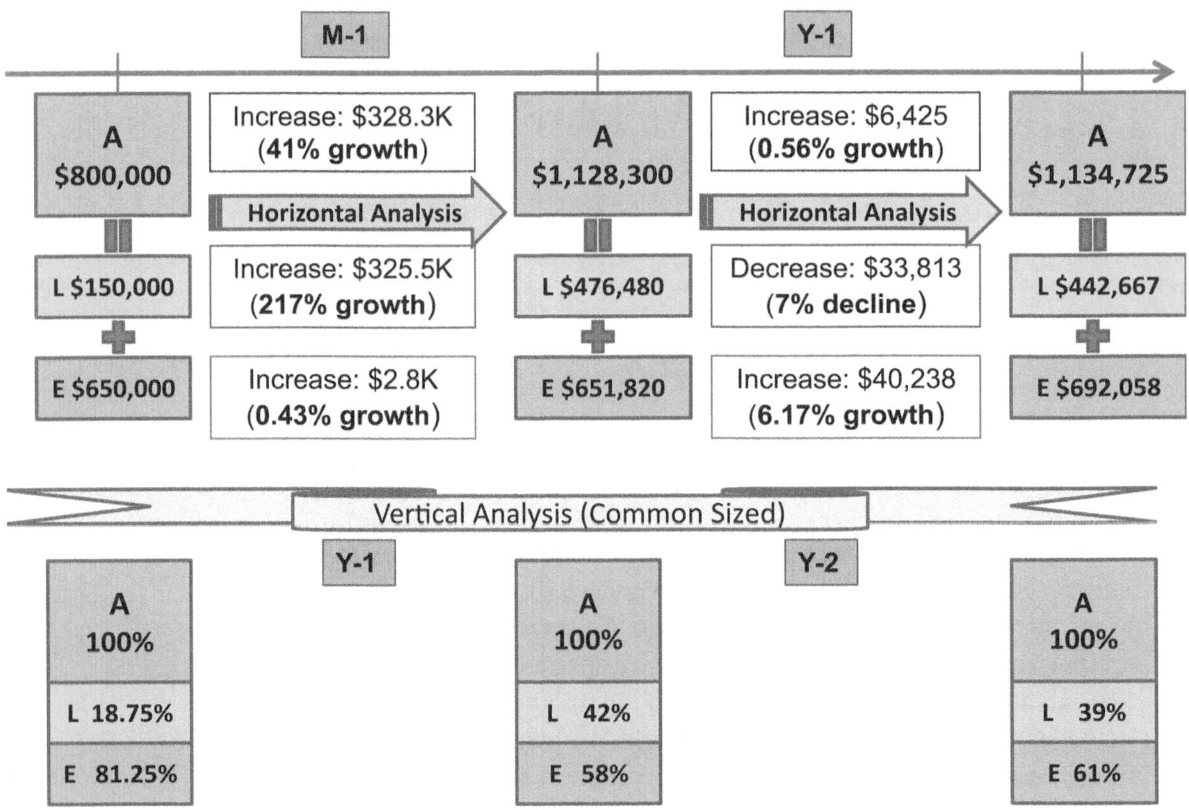

Figure 4.4 Analysis of the balance sheets – R&B Grill.

Figure 4.4 shows the two different ways of viewing the balance sheet records. The top half shows the changes of the amount of each component and presents the changes in percentage. The assets of R&B Grill have grown from \$800,000 in the beginning to \$1,128,300 after 1 month. This is growth by 42%. After 1 year, they grew by 0.56% to \$1,135,705. During the first month, its liabilities grew by \$325,500, which is 217% growth. This huge growth of liabilities was caused by the internal investing (or resource allocation) activities. Then, the next year, the liabilities grew only by 7% to \$442,667. Finally, the equity grew, during the first month from \$650,000 by 0.43% to \$652,800, then it grew during the next year by 6.17% to 692,058. It is clear that this company has grown its assets mostly by taking large amount of debt. The increase of its liabilities shows it clearly. The horizontal analysis this way calculates each component's growth over time by comparing its change. The following formula shows how to calculate the growth percentage (or growth rate):

$$\text{Growth rate (\%)} = ((\text{Current amount} - \text{Previous amount})/(\text{Previous amount})) \times 100$$

The lower half of Figure 4.4 shows the structural changes of the financial position of R&B Grill by vertical comparison. Because the company's total assets are made of liabilities and equity, the total amount of liabilities is divided by that of the total assets to determine the proportion of the liabilities in the total assets. Then, the result is presented in percentage by multiplying 100. The proportion of the equity is determined in the same way. The amount of total equity is divided by that of the total assets and converted to the percentage by multiplying 100. The following is the formula to determine the proportion of each component for a horizontal analysis:

$$\text{Liability percentage (\%) over Total Assets} = (\text{Total Liabilities/Total Assets}) \times 100$$

$$\text{Equity percentage (\%) over Total Assets} = (\text{Total Equity/Total Assets}) \times 100$$

The vertical analysis of R&B Grill's Balance Sheets reveals the changes of its financial positions over time. When it started, its assets were contributed by less than 19% of the liabilities and the rest have been contributed by its owners. The proportion of the liabilities has grown to almost 42% after the first month. Then, it decreased to slightly about 39% after the operations of Y-1.

The same techniques can be applied to the income statements as displayed in the following Figure 4.5 "Horizontal Analysis of the Income Statements – R&B Grill."

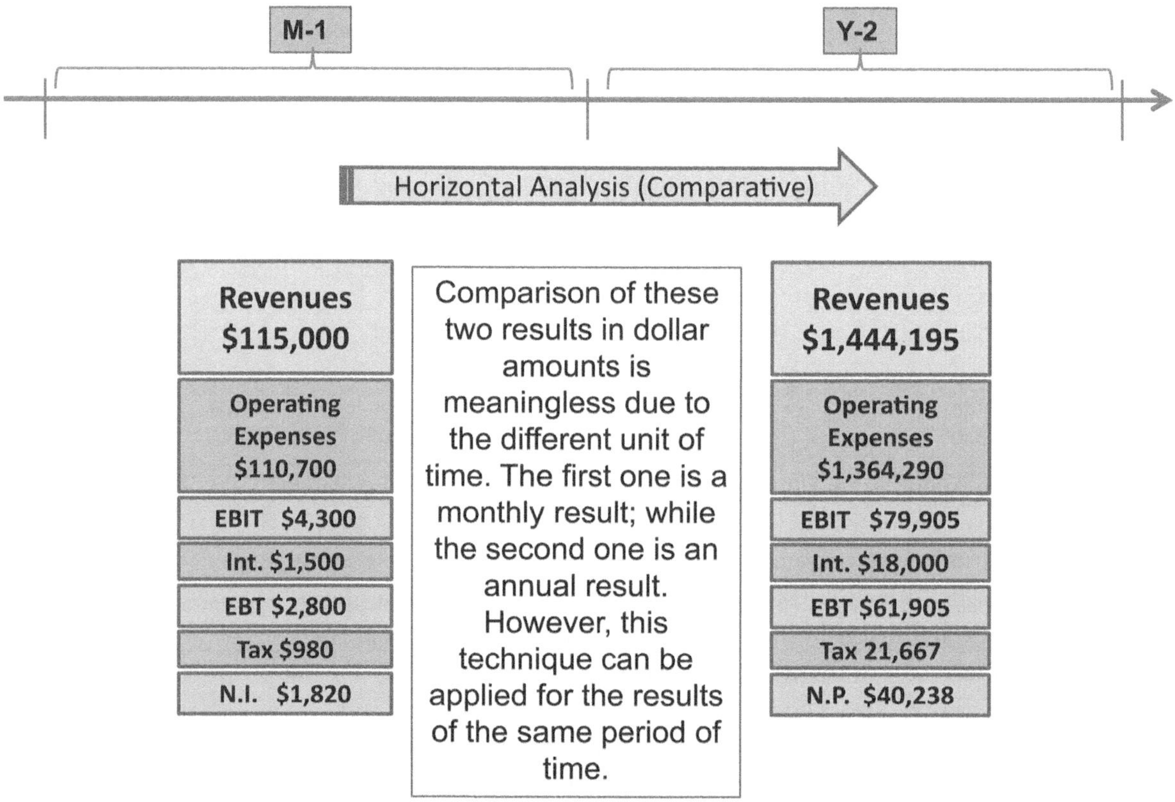

Figure 4.5(A) Horizontal analysis of the income statements – R&B Grill.

The horizontal analysis of R&B Grill's operational results through its Income Statements is meaningless due to the nature of the information presented on the two Income Statements. The first column represents only 1 month's operations during the first month, while the second one covers the entire year after the first month. If, however, the length of the time for the two different operations were the same, this approach will provide valuable information on the changes of the amount of individual items in time. The vertical analysis, which is presented in the following, will provide more meaningful information despite the difference between the lengths of time covered.

The next is the vertical analysis of the Income Statement, which presents the operating efficiency of each time period by showing the percentage (%) of individual expenses and profits over the revenues. The formula that calculates the percentage of each item is as follows:

The percentage of an expense over revenues = (the amount of an expense/revenues) × 100

The percentage of a profit over revenues = (the amount of profit/revenues) × 100

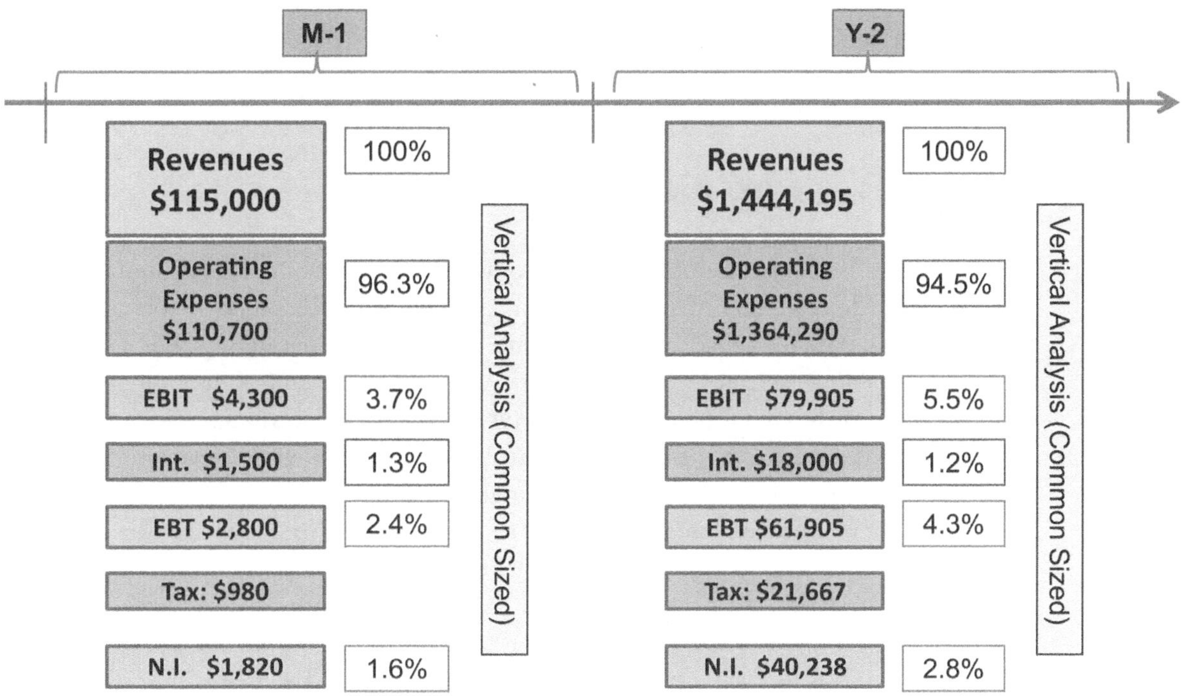

Figure 4.5(B) Vertical analysis of the income statements – R&B Grill.

Although the volume of each item presented on the Income Statements are completely different due to the different time period, this approach can yield useful information regarding R&B Grill's management and operations that were introduced as the "Operational System" earlier. The operational system covers many different features. One of the most important features is the capability controlling the costs – individually and overall. The percentage column indicates that the business spent less in its operating expenses in the second term compared with those of the first. The operating expenses in Figure 4.5(B) includes the cost of sales. The management of R&B Grill apparently saved a lot in the operating activities by reducing expenses by 1.8% (the difference between the operating expenses percentages; 96.3%–94.5%), and it increased the EBIT from 3.7% in the previous term by 1.8% to 5.5% in the second term. Although the 1.5% difference may look small, it represents the 1.8% of the revenues ($1,444,195), which is close to $26,000 (=1.8% x $1,444,195). Thanks to this savings, the Net Income has grown to 2.8% from the 1.6% of the first term. In conclusion, R&B Grill is enhancing its operational efficiency by not wasting its resources in its operating expenses. The final result of Net Income may look disappointing but it must be remembered that R&B Grill just started its business. It will have to grow long into the future.

4.4 RATIO ANALYSIS

The previous sections provided basic concepts and techniques of how to analyze Financial Position and Operating Results by investigating the Balance Sheets and the Income Statements separately. Albeit useful, this approach is limited in that they only address either operational aspects or financial positions separately. For an advanced analysis to develop a comprehensive picture, more specifically designed financial ratios are applied. They are presented in the form of a fraction – one record divided by another. This simple format, however, easily hides the true meaning. Careful interpretations must be made by comparing the two different records (that are the numerator and the denominator used in the formula) to identify the cause (or input) and effect (or output) hidden in different aspects of a business. In other words, ratios, by themselves, do not tell us the real problem. They only help us to ask more questions to identify the true problem. Important is that we need to identify real factors that have caused the business conditions presented in certain ratios. To do this, we must ask such question as "What has caused the changes of a certain ratio over time?" Once these questions are asked, it depends on an individual manager's ability to develop the most reasonable answers to the questions. Developing the most reasonable answers requires in-depth analysis. For this reason, we have to incorporate necessary techniques on our own intelligence by addressing all necessary dimensions of a business. Only by doing so, we can come up with reliable answers. More explanation is provided in the following sections.

4.4.1 Major Ratio Groups

Many textbooks on Managerial Accounting and Financial Management introduce ratio analysis. Titles may differ from one book to another, but their nature and interpretations remain the same. The following is the most commonly introduced group titles of ratios:

1. Liquidity ratios
2. Solvency (or Leverage) ratios
3. Activity ratios (some books name this "Turnover ratios.")
4. Profitability ratios
5. Operating ratios (Some books name this "Activity ratios." Please do not confuse this with the "Turnover ratios" presented above.)

The liquidity ratios contain a few different ratios but the common nature of them is that they all reveal the business's ability of handling its short-term obligations. In plain language, all liquidity ratios point out whether the business can pay its short-term debt with its currently available resources. The solvency ratios are about the long-term debt payment capability. The activity ratios (which are sometimes called turnover ratios) reflect the capability of the management to utilize its resources to generate revenues. This – comparing the resources of a firm with its operating results – is a very important point of view for readers to develop to understand frequently ignored aspects of a business in analysis. The profitability ratios measure the management's ability to generate profits from using its resources and/or from its sales by controlling its costs and expenses. Finally, the operating ratios are supplementary information obtained from operating activities of serving customers. Due to this nature, some books introduce these as "Activity ratios."

Adopting the contents of a few books, this book presents six different groups of ratios as followed by adding the Market Value ratios. They are frequently used in the financial market to assess a firm's performance.

Each group of ratios addresses specific dimensions of a business as explained above. It must be kept in mind that titles may slightly differ from one book to another. As long as readers understand the meaning of individual ratios, the different names of the same ratio can be easily understood. Formulas are introduced later in this section. Readers are encouraged to memorize them with understanding meaning. The following is the summary of the ratio groups with commonly used ratios in each group.

1. **Liquidity ratios** measure how easily the firm can pay its **short-term debt**.

   Examples: Current Ratio, Acid-Test Ratio, etc.

2. **Solvency *(Leverage)* ratios** show how heavily the company is in debt **(long-term debt)**.

   Examples: Leverage Ratio, Debt-to-Equity Ratio, Number of Times Interest Earned Ratio, etc.

3. **Activity *(Turnover)* ratios** measure how productively the firm uses its assets (to generate revenues).

   Examples: Asset Turnover Ratio, Fixed Asset Turnover Ratio, Inventory Turnover Ratio, etc.

4. **Profitability ratios** measure the firm's return (on its revenues as well as on its investment).

   Examples: ROS, ROA, ROE, EPS, etc.

5. **Market-value *(Investor)* ratios** show the perceived value of a firm in the market.

   Examples: Price per Earnings (P/E) ratio, Market-to-Book ratio, etc.

6. **Operating ratios** these are not financial ratios, but frequently used to address financial performance of a business operations.

   Examples: ADR (Average Daily Rate), Occupancy rate, RevPAR (Revenue Per Available Room), Seat turnover, etc.

From the list introduced above, this book will focus mainly on a few popularly used ratios to present how business performances can be interpreted. R&B Grill's financial records will be continuously used for this purpose. Let's reopen Figure 4.3: Wealth flows of a business – R&B Grill (see Figure 4-03). It is unnecessary to repeat the horizontal and vertical analyses of the financial statements. Instead, let's approach differently to view the entire business dimensions by connecting its balance sheet records to the income statement records.

When business records are presented in a chronological order as in Figure 4.3, individual components of the financial statements can be interpreted as the cause and effect of individual activities of the business. The following statement is an example of how the first period (M-1) of the financial records can be interpreted:

"When the business started, it had $800,000 of assets that were composed of $150,000 of liabilities and $650,000 of equity. During the first month, R&B Grill utilized its assets to generate the revenues of $115,000. Out of the revenues, the operations spent $110,700 for its expenses. In result, the company was left with $4,300 as the profits from its operations. After recording the interests on the loan, the taxable income is $2,800. In the meantime, the assets have grown by 41% from its original amount to $1,128,300. This growth has been achieved mostly by the increase of the liabilities that grew more than three fold. As the result of operations after assessing the Income Tax of $980 at 35% of the EBT, the $1,820 in the net profit increased the owners' equity from $650,000 by 0.28% to $651,820."

The second period (Y-1) can be presented as follows:

"During the second period (Y-1), R&B Grill utilized its assets to generate revenues of $1,444,195. During this period, the operational expenses were $879,990, which was 94.5% of the revenues. This was a better performance compared with the first period (M-1) when 96.3% of the revenues were spent in operational expenses. The annual interest expense was $18,000. This was 1.2% of the revenues, which was down by 0.1% from the previous period of M-1. The result of the operations yielded 4.3% of revenues as the net profit. The total amount was added to the Retained Earnings; and the Owners' Equity increased by 7.95% to $714,705 at the end."

The interpretations clearly present the business performance that ended up in the increase of the owners' equity by introducing not only the dollar amounts but also their percentages. However, they failed to expose specific information such as whether the business is utilizing its investments effectively or whether their operating efficiency improves to provide returns to the investors. Now, it is time to analyze specific aspects of the business by itemizing individual ratios. Once necessary ratios are introduced, they will be combined into a comprehensive analysis of the business.

4.4.2 Profitability Ratios

The ratios in this group measure the profitability of the firm's operations and financial activities.

Figure 4.5, which was used to introduce the vertical analysis of the Income Statement, is reintroduced to point out the most important profitability ratio that represents the management's capability of controlling its costs (see Figure 4-05(B)). It is the Return on Sales (ROS). It is often called the Net Profit Margin or the Net Profitability. The calculation of this ratio is dividing the net profit (or net income) amount by the revenues for the period; and it is presented as a percentage. This ratio during the period of M-1 (in Figure 4.3) is 1.6%. This means that out of the entire revenues, only 1.6% of them were left in the company as its profit. In other words, 98.4% of the total revenues were expenses. Because the profit is the leftover from the revenues after taking out all expenses, it is called the "margin." The next period (Y-1) shows a little enhanced result. The ROS is 2.8%. During the second operating period, the expenses were 97.2% of the revenues, which was down from the 98.4% of the M-1 by 1.2%. Again, the 1.2% difference in the ROS is, in fact, $38,808. This is one important fact to consider when ratios are used to analyze. Always, the base of the percentage must be considered. The following is the summary of the ROS ratio.

ROS (Return on Sales, (Net) Profitability, (Net) Profit Margin) measures the profit percentage of the revenues. This ratio also indicates the overall cost-control capability of the operators.

Formula: (Net Income/Revenues) × 100 = ROS (in %)

After the analysis of the ROS for the operational efficiency in the cost-control activities, another dimension of efficiency can be addressed by focusing on the revenues and total assets. Revenues were generated by utilizing the entire assets of the business. The company owns its assets as the result of its investing activities (as also described as "resource allocation"). As presented earlier, the primary result of the management's investing is the revenues generated using them to serve customers. Only after generating revenues, net profits can be expected. When these relationships are identified among assets, revenues, and net profit, it is easy to connect the relationship between the assets and the net profit. In this view, we are bypassing the revenues by directly relating the net profit as the result of the assets. This result is presented in the ratio called the Return on Assets (ROA). Just like the ROS, the ROA is the percentage of the Net Income of the amount of Total Assets used. However, caution must be practiced in this analysis. The amount of total assets constantly changes over time as we have seen in previous chapters because revenues increase assets and expenses decrease assets. For this reason, we need to calculate the average daily amount of total assets as follows to compare it with the revenues:

Average Assets = (Beginning Assets + Ending Assets)/2) → Average amount of daily assets used

Once the average amount of assets is determined, the ROA can be calculated as follows:

ROA Formula: (Net income/Average Assets) × 100 = <u>ROA</u> (in %)

As shown in Figure 4.3, there is a bent arrow connecting the assets of each term to the revenues. During the M-1, the assets have grown from $800,000 to $1,128,300. Simply, R&B Grill has increased its asset amount rapidly. So, the average amount of daily assets it used is $964,150 (← ($800,000 + $1,128,300)/2)) during M-1.

By using the average assets of $964,150 every day, the company has generated $1,820 in its Net Income in the M-1. So, its ROA during M-1 is 0.18%. Of course, this is a very small return; but it was also only for the first month of operations. The ROA of the next term (Y-1), when calculated the same way by dividing the net income ($40,238) by the average asset $1,113,514) is 3.6%. This business has obtained 3.6% of its invested amount as its net return. When the assets are considered as the company's investment, the ROA becomes the company's Return on its Investment (ROI). This is an important concept for operators and investors together to understand. The ROI of the investors must be assessed differently.

As repeatedly pointed out, a company's assets are contributed by two groups of stakeholders that are creditors and the investors. This structure is well represented in the fundamental accounting equation of Assets = Liabilities + Equity. Under the structure, the owners' investment is considered the Equity of the company and the management adds the net income back to it. In this context, the net income returned by the management to owners is the investors' ROI. Just like the assets, the amount of owners' equity inside the company changes constantly because revenues increase equity and expenses decrease it. For this reason, when the net income is compared with the equity to determine the rate, the average equity amount must be used. The result of this calculation of dividing the net income by the average equity is called the Return on Equity (ROE). The following is the summary of technical details of the ROE.

Average Equity = (Beginning Equity + Ending Equity)/2) → Average amount of daily equity

ROE Formula: (Net income/Average Owners' Equity) × 100 = ROE (in %)

The reason that the net income is returned to the equity is that the creditors of the firm already have received their return before the net profit was determined. Their return (or compensation) is the interest expenses presented in the Income Statement before the net profit.

Figure 4.3 shows the ROE, which is the ROI of the investors. After the first month's operations, the equity grew by $1,820. According to the business transactions processed in previous chapters, there was no additional equity investment made since the original. Applying the formula, the ROE of the first month is 0.027%. This is very small, but again, it is just the result of only the first month. The next year (Y-1) has yielded almost 6% of ROE. The investors have received close to 6 cents of return on their investment of $1. The following is the calculation process of the ROE of the Y-1.

(Net Profit of the Y-1 divided by the Average Equity of Y-1) x 100 = ROE

$40,238/(($651,820 + $692,058)/2) x 100 = 5.98%

In summary, R&B Grill has conducted its business during the Y-1 and achieved the ROS of 2.8%, the ROA of 3.6%, and finally the ROE of 6%. These are the profitability ratios. There are other similar ratios in this group; but these three are most widely used in business analysis. All others use specific types of profits (such as Gross Profit, EBIT, EBT, and Net Income) with more narrowed down items to pinpoint a unique condition of a business. Once these three ratios are completely understood, the same concepts can be easily applied to other profitability ratios to identify specific dimensions of the operation. The next topic is the Activity Ratios (or Turnover Ratios).

4.4.3 Activity Ratios (or Turnover Ratios)

Ratios in this group measure the management's effectiveness of using its resources in generating revenues.

Ratios that belong to this group show how effectively a firm uses its resources (that are assets) to generate revenues. A company must use its assets (properties and other necessary resources) to serve its customers, and only by serving customers, it generates revenues. When we think of this sequence of the flow of wealth, the connection from assets to revenues can be explained as input→output relationship. It is presented by the bent arrows in Figure 4.1 from assets to revenues.

4.4.3.1 Asset Turnover Ratio

Once we understand that using assets (to serve customers) is a necessary condition to generate revenues, the input–output relationship between the two becomes clear. The management of a company must invest to obtain the most effective assets and must generate the largest amount of revenues using them. Figure 4.1 depicts this process. The necessary assets are put into the operational system that will generate revenues as output. In this context, we can determine the numerical value to measure the effectiveness by dividing Revenues (output) amount by the Total Asset amount (input). In this application, as the amount assets always changes as the revenues are generated, the average amount of total assets must be used in this calculation as in the case of ROA and ROE.

Average Assets = (Beginning Assets + Ending Assets)/2

Formula: (Revenues/Average Assets) = Asset Turnover (# of times)

Let's use the first year (Y-1) of R&B Grill to explain this ratio. The total assets of $1,128,300 in the beginning of the Y-1 have grown to $1,134,725. Every day during Y-1, R&B Grill generated total revenues of $1,444,195 using the average assets of $1,131,512.50. Its assets turnover was 1.28.

In business terms, this means that the company's entire invested resources (which is average assets) have been turned over (which means that they were used or utilized) 1.28 times during the year, or the company's revenues (as output) were 1.28 times as large as its average assets (as input). It can also be said that one dollar ($1) invested into its assets (as input) has generated $1.28 in its revenues (as output). The logic of this ratio indicates that good investments into useful resources will generate more revenues. This also means that if this ratio turns out poor in comparison with others, the investment into its assets might be a poor decision. This ratio is frequently used to determine the effectiveness of the managements' investing decisions.

4.4.3.2 Inventory Turnover Ratio

Another important operational efficiency of a restaurant business is measured by the Inventory Turnover ratio. Inventory control is unquestionably critical in restaurant operations. Keeping the right amount of inventory is the key to running the operations smoothly. Keeping too much inventory will reduce the available cash due to the excessive purchases. Too much inventory on hand will end up in wastage eventually. Keeping very small amount of inventory may result in shortage of ingredients, which will eventually cause lac of sales by not being able to serve customers. For these reasons, operators must determine the right amount of inventory to keep, the right time to place orders, right amount to order to maintain the best efficiency.

Regarding inventory control, by comparing the average inventory amount on hand with the amount used in serving customers, it is possible to identify important operational characteristics of a business. The average amount of inventory on hand is determined the same way as the average assets or equity was determined. Simply, it is the average between the beginning inventory and the ending inventory of each accounting period. There must be some fluctuations during one accounting period, but if a business is reasonably managed, it must maintain relatively a stable amount of inventory on hand every day. The amount of inventory used is calculated at the end of each accounting period to determine the Cost of Sales. Using the two pieces of information – average inventory and the Cost of Sale – we can identify how often the business has used up its entire inventory and replenished it during the period. It is called "Inventory Turnover" ratio. The following is the formula of the ratio.

4.4.3.3 Inventory Turnover Ratio: Cost of Goods Sold (COS)/ Average Inventory

A higher inventory turnover indicates that the business replenishes its inventory more often. If this ratio is very low, the business may be keeping too much inventory on hand. Too much inventory on hand indicates the business has spent its cash inefficiently by purchasing too much. If it is too high, on the other hand, it may indicate an ineffective control system that causes frequent shortages of necessary merchandise. In this case, the business loses sales.

R&B Grill's average inventory amount on hand during Y-1 is $33,650 (the average of the beginning inventory amount of $34,800, and the ending amount of $32,500). The Cost of sales (COS) of Y-1 is $484,300. When the amount of the COS is divided by the average inventory amount, the inventory turnover ratio is 14.4, which means that the business has completely replenished its entire inventory 14.4 times during the year. It is a usual perception that a restaurant with a high inventory turnover serves fresher products. However, too high inventory turnover ratios may be an indication of such problem as purchasing too little inventory at a time. As the result, the business may experience frequent shortage of its inventory supplies.

This ratio can be extended further to identify another characteristic of a firm's operations. By dividing the Inventory Turnover ratio into the number of the days of a year, we can determine how many days the business uses its inventory on hand. The following is the formula.

365 days/Inventory Turnover Ratio = the number of days using one batch of inventory

When this approach is applied to many successful restaurant companies, their inventory turnover takes about 2 weeks. Thus, besides the financial impacts caused by the inventory control system, this ratio can be used to estimate the freshness of the products, which is qualitative side of the operation. If one batch of inventory purchased is stored too long inside the business, it will lose freshness. On the other hand, too small number also indicate problematic situation. If a business has to order its inventory too often, it must allocate dedicated personnel to receive it correctly. The preparation of them such as cleaning and processing may become very difficult to be thorough. Because food preparation must be conducted with caution, the operators must carefully design its inventory control plans to avoid unnecessary losses.

Dividing R&B Grill's inventory turnover ratio of 14.4 into 365 days, the result is 25.3. This means that R&B Grill used its average inventory on hand for about 25 days. This information becomes more useful when compared with other similar operations. Chipotle, one of the most successful restaurant companies in this dimension at the time of writing this script, for example, shows its inventory turnover ratio at 99 times during the 2015 fiscal year. It means that the average amount of inventory on hand lasted only 3.7 days. Consequently, Chipotle must purchase its inventory almost twice a week. This is an extremely efficient inventory control, only if the business can sustain it with no negative side effect. Another successful company, Bloomin' Brands, which is well known for its flagship concept of "Outback Steakhouse", turned over its inventory 17.6 times during the same fiscal year. Its average inventory on hand lasted 21.7 days during the year.

From this comparison, R&B Grill's inventory control policy may need to improve. However, these results should not be used to judge that one company's practice is better than the other. It depends on the service concept and the nature of the products of individual businesses. However, it may help them achieve better efficiency using other successful companies' ratios as a benchmark.

4.4.3 Advanced View of Ratio Analysis

4.4.3.1 ROA Disaggregation

There is a very important, and also interesting, approach in using ratios to analyze the performance of a business. The profitability ratios –

ROS, ROA, and ROE – use the net profit as the common factor to measure the capability of a business to generate return from three different dimensional activities. ROS ratios measure the efficiency of operating cost-control dimension. ROA ratios address the efficiency of the investing activities of the business itself. Finally, ROE ratios measure the return on the net investment made by the owners, which is the ultimate goal of a business. The interrelationship of these three dimensions can be addressed from the following perspective:

First, a firm must invest its capital financed through loans and equity as effectively as possible to generate the maximum amount of revenues. This ability is presented in its Asset Turnover ratio of each year. Once investing activities are correctly performed, the next step is to maintain its operational efficiency by controlling costs within acceptable range. When the control system works smoothly, the ROS will be maintained at its highest level as possible. When these two dimensional activities are correctly conducted, the company will achieve the highest ROA ratio. These sequential activities are presented in the formula called the ROA Disaggregation. It is introduced below.

1st dimension → 2nd dimension → Company's ROI

Asset Turnover × Return on Sales = Return on Assets

The Asset Turnover Disaggregation formula is not about the simple application of algebra. Instead, it stands for the function of two different dimensions of a business. First, managers of a business must make good decisions on its investing activities not to waist resources; then, it must control its costs in operating activities thoroughly. Once these two dimensional activities are effectively and efficiently conducted, the result will be the highest return on its investments (ROI) of the business.

4.4.3.2 ROE Disaggregation

It must be reiterated that the ultimate goal of running a business is always to provide the maximum return to its equity investors. To achieve this final goal, there is an additional dimension a business must face. Since the entire amount of net profit becomes the return to the owners, the rate of return on the equity investors is the ratio between net income and the amount of equity as presented by the ROE ratio. Naturally, smaller amount of equity will raise the ROE higher. This means that if a business keeps higher amount of debt, it will be able to provide higher return to the equity investors. In other words, the financing decision with more debt may be a better decision for the equity investors. Just like the ROA Disaggregation formula introduced above, another sequential approach is possible to present this effect.

$$\text{Company's ROI} \times \text{Financing Decision} = \text{Investors' ROI}$$

$$\text{Return on Assets (ROA)} \times \text{Leverage} = \text{Return on Equity}$$

Once the management of a company conducts its primary tasks of effective investing and thorough cost control to achieve its highest return on its investments, the next dimension is to determine the best mix of its debt and equity. By taking this extra step of determining the best capital structure, the business can finally provide the highest return to the equity holders for their investments. These sequential activities of a business can be extended in the following equation.

Activity or decision-making dimensions of the management of a business firm

$$1^{st}\text{ dimension} \times 2^{nd}\text{ dimension} \times 3^{rd}\text{ dimension} = \text{ultimate goal of a business}$$

$$\text{Asset Turnover} \times \text{Return on Sales} \times \text{Leverage} = \text{Return on Equity}$$

Although the sequence is presented in the form of an algebraic equation, the real meaning of this approach is not in calculation but in pointing out the importance of business decisions that emphasize specific dimensions of a business that are investing activities, operational efficiency, and finally the financing decisions to determine the most efficient structure of the capital. When a business is smoothly running with its highest efficiency in all these dimensions, it finally ends up with maximized return to the equity investors by satisfying all other stakeholders in the middle.

Let's use the example of R&B Grill to explain the ROA Disaggregation. During Y-1, it has turned over its entire assets 1.28 times. In result, it has generated the revenues of $1,444,195. From the revenues, it has earned the net profit of $40,238. Its ROS was 2.8% in ROS. When the

ROA is calculated using the formula, it was 3.6%. Let's use the Asset Turnover Ratios and the ROS in the sequence introduced earlier.

Dimensions: (Investing) × (cost control) = a firm's ROI (performance)

Ratios used: Asset Turnover × ROS = ROA

1.28 times × 2.8% = 3.6% (rounded)

R&B Grill's entire investment has been used (or turned over) 1.28 times to generate revenues; and the cost-control effort has finally brought 3.6% of return on its investments. Rachel and Brad must analyze their business results from this perspective continuously to identify which dimension is their strength. By doing so, they can establish most appropriate policies to achieve the best results. ROE Disaggregation will be explained in Chapter 10 with the topic of Valuation.

Although there are a lot of other ratios that can pinpoint specific aspects of business performance, this chapter presented only a few major ratios with their applications using R&B Grill's financial information. Readers are encouraged to apply the concepts and techniques to other businesses.

CHAPTER 5

COST ANALYSIS AND CONTROL

5.1 INTRODUCTION

This chapter introduces the nature of costs in general and provides a simplified method that can be used to control them in a restaurant business. As the term "cost" is casually used to represent spending for various purposes, controlling cost could be a complex issue that permeates the entire aspects of a foodservice business. For example, the term "cost" often means spending (or expenditure) for internal investment, such as purchasing a piece of equipment or furniture. It may also indicate lost revenues which otherwise could have been earned in a statement such as "Bad weather of last winter has cost us at least $10,000." Although they are casually mentioned this way, cost has its specific nature as introduced in this chapter.

The term "cost" in this chapter refers to the spending to generate revenues. In this context, the nature of cost is directly or indirectly related to the revenues. For example, food costs and labor expenses can be considered direct costs, while expenses for utility, rent, or insurance can be considered indirect costs. These types of spending are presented on the Income Statement as expenses. Instead of covering a variety of views on different types of costs as introduced in many textbooks, this book will limit its focus on the nature of costs in relation to the sales volume and will provide a framework to control them to achieve the financial goal of a restaurant business. All restaurant businesses present their Income Statements with the revenues on the top followed by cost of sales, labor costs, other operating expenses, interests, and finally tax expenses before the net profit.

Previous chapters have introduced how accounting information is generated and presented by processing the data collected from a variety of business activities. In Chapter 4, the final sets of information presented on each statement were compared and analyzed to show how to view the performance of R&B Grill. A simple vertical analysis of R&B Grill's Income Statement shows that the profitability of 2.8% in the past year was a better result than its previous term's performance of 1.6%. Although the operational performance has improved over time, it is not known whether the progress was good enough. The cost of sales, for example, has declined from 39.3% in the first month to 33.5% in the next year. This reduction may be taken as improvement, but it does not tell whether the quality of the product was maintained as planned. The same is true with the labor expense. It declined from 33% in the first month to 27.8% in the next year. It may be considered a good sign that the labor productivity is improving. However, the result does not tell whether the savings were truly the result of the efficiency improvement. It is possible that excessive wage control may cause grievance to employees, and it may hurt the quality of service to customers. In summary, costs must be carefully analyzed and controlled with the quality of the product and services taken into consideration. This point of view on cost control leads to a notion that there must be a predetermined amount of actual costs when a certain amount of sales is projected. This is the concept of standard costs. Let's start investigating the standard costs with the prime costs of a restaurant business.

5.2 FOOD AND LABOR COSTS

Food and labor costs were introduced earlier as the "Prime Costs" of a restaurant business. It is commonly said that "prime costs must be controlled within two thirds of the revenues." This statement, as a rule of thumb, must reflect the generally accepted experiential learning of restaurant business practitioners. Let's apply the case of R&B Grill (refer Table 3.6: The Income Statement in a Revised Format).

According to Table 3.6, the food cost (the Cost of Sales) in relation to the revenues has declined from 39.3% to 33.5%. Its labor cost also declined from 33% to 27.8%. Combining these two prime costs, R&B Grill has spent 72.3% of its revenues on prime costs in the first month, which was above the rule of thumb of the two-thirds of the revenues. In the next year, its prime costs were controlled down to 61.3%, which was more than 5% below the rule of thumb (66.7%). As mentioned earlier, however, it is not known whether this reduction is a good sign or not. There must be a benchmark that can be compared with the result. Only then, we can conclude whether the business is truly improving its efficiency. Let's find out how individual costs – food cost and labor cost – can be standardized into ideal ones.

5.3 THE STANDARD FOOD COST

Food cost is simply the amount of food and beverage inventory used in serving customers. Restaurant managers can develop their system to pinpoint how much inventory they should use in operations by developing good inventory control system. Many successful restaurant businesses separate their inventories into subcategories. Breaking down inventories as below helps practitioners pinpoint how much they are spending on specific items or groups of inventory:

Bakery, Dessert, Coffee, and Tea (other drink), Dairy, Dry (canned or bottled included), Meat (frozen and fresh), Poultry, Produce, Seafood, etc.

The standard food cost can be determined by developing the standard recipe of each menu item and by recording the price of each items as well. The details of this approach will be provided with an imaginary list of menu items with their standard costs presented in Table 5.1.

Table 5-01 Menu Items and Cost Structure

Menu Item	Menu Price	Standard Cost	Ideal Cost %
Appetizer item A	$ 6.00	$ 2.25	37.50%
Appetizer item B	$ 8.00	$ 3.45	43.13%
Drink item A	$ 4.00	$ 0.85	21.25%
Drink item B	$ 5.00	$ 1.25	25.00%
Entrée item A	$18.00	$ 6.50	36.11%
Entrée item B	$24.00	$ 8.50	35.42%
Dessert item A	$ 7.50	$ 2.25	30.00%
Dessert item B	$ 9.50	$ 3.25	34.21%
Alcoholic item A	$ 5.00	$ 1.25	25.00%
Alcoholic item B	$ 7.00	$ 1.75	25.00%

Table 5.1 shows the list of menu items of R&B Grill with the price and the standard cost of each item. Restaurateurs must develop their standard recipe and the cost of each menu item. According to the information on the table, if one (1) of the Appetizer Item A was sold, the revenues must be $6.00 and the ideal food cost for the item must be $2.25, which is 37.5% of the revenue. To use this type of framework, it is important that the cost of each item must be adjusted periodically as the market conditions change. Of course, the menu price may be adjusted when necessary. This is how the ideal cost can be estimated in the beginning but it is not sufficient yet. There are additional sets of necessary information that must be obtained from the operating results. It is the operators' responsibility to create effective record-keeping systems to accumulate operating information such as the customer counts, the number of individual menu items sold, the records of employees' working hours, to name only a few. Table 5.2 presents the summary of R&B Grill's operating results of the sales mix and the ideal costs during the second accounting period (Y-1). Explanation of the details is provided under the table.

Table 5-02 Sales Mix and Ideal Food Cost %

Menu	Menu Price	Standard Cost	Ideal Cost %	Order (%)	Per Capita Spending	Ideal Cost *
Appetizer item A	$ 6.00	$ 2.25	37.50%	2.00%	$ 0.12	$ 0.05
Appetizer item B	$ 8.00	$ 3.45	43.13%	4.00%	$ 0.32	$ 0.14
Drink item A	$ 4.00	$ 0.85	21.25%	36.00%	$ 1.44	$ 0.31
Drink item B	$ 5.00	$ 1.25	25.00%	46.00%	$ 2.30	$ 0.58
Entrée item A	$18.00	$ 6.50	36.11%	48.00%	$ 8.64	$ 3.12
Entrée item B	$24.00	$ 8.50	35.42%	38.00%	$ 9.12	$ 3.23
Dessert item A	$ 7.50	$ 2.25	30.00%	4.00%	$ 0.30	$ 0.09
Dessert item B	$ 9.50	$ 3.25	34.21%	6.00%	$ 0.57	$ 0.20
Alcoholic item A	$ 5.00	$ 1.25	25.00%	32.00%	$ 1.60	$ 0.40
Alcoholic item B	$ 7.00	$ 1.75	25.00%	36.00%	$ 2.52	$ 0.63
Total				252.00%	$ 26.93	$ 8.73
Total Number of Customers served: 53,650					Ideal Cost %	32.41%

*The Total of Ideal Cost is presented in rounded amount to the nearest cent. Because of the rounding effect, the total amount may be off by 2 cents.

The first three columns from the left are identical to the Table 5.1. They are individual menu items, their prices and the standard costs estimated from the standard recipe. The next column that says "Order (%)" shows the percentage of each item ordered by customers during the accounting period of Y-1. Throughout the year, R&B Grill served 53,560 customers as presented on the bottom part of the table. This information must be made available for reliable analyses. The records in the Order (%) column indicate that the 2% of the 53,560 customers have ordered the "Appetizer item A," 4% of them ordered "Appetizer item B," and so forth. The bottom line in the column says 252.00% in total. This means that one customer, on average, ordered 2.52 items while dining at R&B Grill. It is easily conceivable that customers usually order multiple items from the menu at a restaurant. The next column (Per Capita Spending) is the average amount spent by each customer on individual menu items. The 2.0% on the first line in the Order (%) for "Appetizer item A" means that 2% of the entire customers have ordered the item during the year (Y-1). Then, the percentage of the order, which is 2% in this case, is multiplied to the Price of the item to calculate the Per Capita Spending on this item. The result is $0.12. This indicates that the average customer has spent $0.12 on the Appetizer item A. Other items can be calculated the same way. When all the amounts of "Per Capital Spending" are added, it equals $26.93. This is the Average Check Price. Restaurateurs usually divide the total revenues (that is $1,444,795 in our example) by their customer counts (which is 53,650) to obtain this information. This simple method is easy to apply, but it does not show the breakdowns of menu mix. The method introduced above provides better information with more details by showing how the revenues are composed.

Using the same concept, multiply the amount of standard cost to the order (%) to determine the ideal cost of each item sold and add up to find out the ideal cost of the revenues per person. The total of the ideal cost of serving one customer turned out $8.73 when the average check price (per capita spending) is $26.93. The ideal cost percentage in this structure becomes 32.4% as presented on the last line of the table. It must be pointed out that the Ideal Cost on Table 5.2 is calculated by the Excel Spreadsheet. It rounded individual costs to the nearest cent. In this process, a few items were presented with slightly higher in their Ideal Costs. If individual costs are added up manually, it may be $8.75. This book will use $8.73 for analysis and projection purposes.

This standard cost percentage can be compared with the actual outcome of the operations on the Income Statement. R&B Grill's Income Statement shows its cost of sales was $484,300 for the year of Y-1, which is 33.5% of the revenues (of $1,444,795). Although the actual cost percentage of 33.5% is higher than the ideal cost percentage (32.4%), it is considered relatively good for a full-service restaurant compared with several national chains. This will be explained later in the book. When the labor cost is added to this, the prime costs took only 61.3%. This result looks very good according to the rule of thumb on the prime cost percentage (67%). However, more in-depth analysis with the ideal cost presented on Table 5.2 indicates that R&B Grill's cost control could improve at least a little further.

Although this approach allows operators to identify specific discrepancies between the ideal food cost and the actual spending, it must be remembered that the actual cost is spent on a variety of items in large volume. The purchase cost of individual items may change in different seasons, which will cause uncontrollable deviation in the cost-control activities. For this reason, the average amount of the Cost of Sales must be used for an accounting period like a year. In addition, the sales mix also is the determinant of the ideal cost. If items with high cost are sold dominantly, the ideal cost percentage will change accordingly. This is why management must have a good record keeping system to obtain necessary operating data.

5.4 FOOD COST AS A VARIABLE COST

As discussed, the food cost can be presented as a percentage of the sales. This is possible because the inventory is consumed only when customers are served. If no customers are served and no revenues are generated, there should be no food cost. The amount of the food costs will increase at the same rate of the cost percentage as more revenues are generated. If R&B Grill's result of the first year (Y-1) is considered ideal, the management can expect its future food costs to be around 33.5% of the revenues. This effect can be presented in a straight line between the food cost amount and the revenues amount as shown in the following Table 5.3 and the graph next to it. The amounts of revenues in the table are arbitrary.

Table 5-03 The Relationship between Revenues and Food Cost

Acct. Period'	Revenues	Cost (33.5%)
1	150,000	50,250
2	200,000	67,000
3	250,000	83,750
4	300,000	100,500
5	350,000	117,250
6	400,000	134,000

(The revenues are arbitrary.)

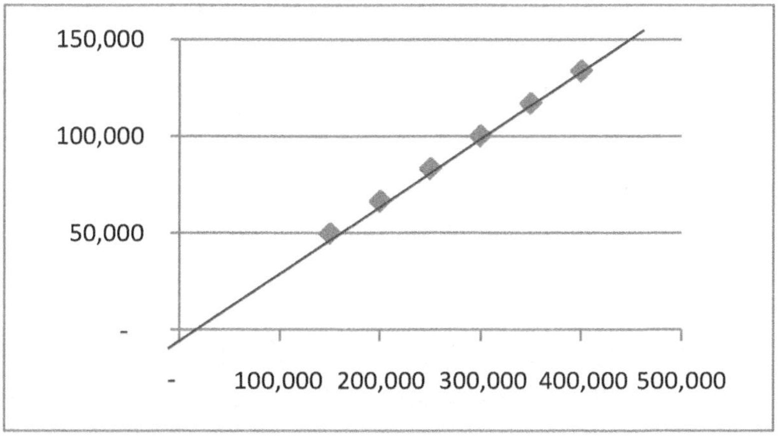

The horizontal axis of the graph indicates the revenues amount of each accounting period, while the vertical axis represents the food cost at 33.5% of the revenues. It is clear that the costs are increasing consistently as the revenues grow and they form a straight line in the graph. When two variables are presented in a straight line like this, it is described that they are linearly related. In this example, the food cost is linearly related to the revenues. The same table and the graph can be prepared by replacing the revenues with the customer counts. Another fact that can be easily noticed is that when there is no revenue, the food cost will be zero (0), too. These attributes help us conclude that the food cost amount can be determined by multiplying 33.5% (which is the ideal food cost percentage) to the sales amount forecast. This is expressed in an equation below:

Ideal Food Cost amount = 33.5% × Expected Revenues

When the nature of a cost has these attributes, it is called a variable cost (VC). A VC changes as the revenues change. However, its percentage of the revenues remains the same. For this reason, we can use the equation introduced above to estimate the ideal food cost amount at any projected revenues. In restaurant operations, the food cost is the only true VC presented on the Income Statement. The food cost must be zero when there are no revenues and it must increase at the same rate of the ideal cost percentage as the revenues grow. All other costs behave differently. To explain this, the other type of cost must be discussed later in this chapter.

In practice, however, real-life situations usually do not work exactly as theory predicts. The food cost fluctuates depending on the market conditions of individual products. Taking into consideration the possibility of changing prices of inventory items, practitioners must absorb certain degrees of variation from the standard cost percentage; and try to maintain the actual food cost close to the long-term target. Let us discuss this with the monthly records of R&B Grill as presented in Table 5.4.

Table 5-04 R&B Grill's Monthly Sales and Food Costs

Month	Customer Count	Revenues	Avg. Check Price	Food Costs	Food Cost %
1	3,750	$ 97,600	$ 26.03	$ 36,500	37.4%
2	4,200	$ 115,600	$ 27.52	$ 40,300	34.9%
3	4,450	$ 122,375	$ 27.50	$ 39,500	32.3%
4	4,850	$ 133,475	$ 27.52	$ 42,500	31.8%
5	4,980	$ 128,335	$ 25.77	$ 43,500	33.9%
6	5,375	$ 150,230	$ 27.95	$ 46,500	31.0%
7	4,350	$ 107,160	$ 24.63	$ 40,500	37.8%
8	3,975	$ 100,370	$ 25.25	$ 34,000	33.9%
9	4,050	$ 108,440	$ 26.78	$ 36,500	33.7%
10	4,300	$ 118,250	$ 27.50	$ 38,000	32.1%
11	4,620	$ 129,360	$ 28.00	$ 42,000	32.5%
12	4,750	$ 133,000	$ 28.00	$ 44,500	33.5%
Total	53,650	$ 1,444,195	26.92	$ 484,300	33.5%
Lowest	3,750	97,600	$ 24.63	34,000	31.0%
Highest	5,375	150,230	$ 28.00	46,500	37.8%

As shown in Table 5.4, the monthly Food Cost percentages (%) are different every month. The customer counts fluctuate and so do the amounts of sales and food costs. The average check price (per capita spending, as introduced earlier in this chapter) and the food cost percentage are different in each month as well. The average check price of each month ranges from around $25 to $28 and the food cost percentage moves between 31% and 37.8%. This is natural due to the seasonality of a restaurant business. Restaurateurs must absorb variance like this and try their best to achieve the long-term goal of maximizing profits by applying operational strategies and tactics. When the entire year's records are compressed to the average, the average check price becomes $26.92 and the average food cost becomes 33.5% as introduced earlier.

5.5 THE STANDARD LABOR COSTS

Along with the food costs, the labor costs belong to the prime costs of a restaurant business. The labor costs also change according to the volume of sales. Higher revenues incur more labor costs to provide necessary services to the customers. The nature of the labor costs, however, is different from the food costs because they are usually made of two different types of costs. One part is a fixed amount that must be spent regardless of the volume of the sales. Salaries expenses are the typical example of this fixed amount. The other is the hourly wage expenses that are considered to change proportionately with the sales volume, which makes them a VC. Although this view of wage expenses seems reasonable, it requires more careful attention. At the time of scheduling employees' working hours based on specific forecast, the wage expenses become VCs. However, it will be difficult to change the work schedule of hourly employees during actual operations. It may be altered to a certain degree but cutting the working hours of employees to reduce the labor expenses may cause grievance of employees. It may result in unwanted high turnover of the employees because they may not be able to make living on such precarious policy. Restaurateurs must design their organization structure very carefully, along with accurate forecasting to maintain the labor costs under control. Sales forecasts and labor schedule must be compared with the actual outcome periodically; and the causes of discrepancy must be addressed to narrow down the gap in the future. In this context, the labor schedule becomes the base of the standard labor costs of each operating period. With the mixed nature that combines the fixed costs (FCs) and the VCs, the labor costs belong to the category of mixed costs (MCs). These will be explained further in the following sections. More details of the labor costs will be explained in the next chapter when forecasting is introduced.

5.6 FIXED COSTS

Different from the VCs, some costs do not become zero when there are no sales. The cost items of this nature can be divided into two separate groups. The first group is the FCs as briefly introduced in the labor costs. The amounts of these costs (FCs) do not change regardless of the revenues. Good examples used in the case of R&B Grill include expenses such as rent, interest, and depreciation, to name a few. Even when revenues are zero, these expenses must be spent. The term "fixed costs" represents the constant nature in their amounts. Apparently, due to their constant amount, their percentages on revenues change over time. When the revenues grow larger, their percentages will get smaller and vice versa.

Table 5-05 R&B Grill's Fixed Costs

Month	Customer	Revenues	Rent	Rent %	Dep.	Dep. %	Interest	Int. %
1	3,750	$ 97,600	$ 15,000	15.4%	$ 6,000	6.1%	$ 1,500	1.5%
2	4,200	$ 115,600	$ 15,000	13.0%	$ 6,000	5.2%	$ 1,500	1.3%
3	4,450	$ 122,375	$ 15,000	12.3%	$ 6,000	4.9%	$ 1,500	1.2%
4	4,850	$ 133,475	$ 15,000	11.2%	$ 6,000	4.5%	$ 1,500	1.1%
5	4,980	$ 128,335	$ 15,000	11.7%	$ 6,000	4.7%	$ 1,500	1.2%
6	5,375	$ 150,230	$ 15,000	10.0%	$ 6,000	4.0%	$ 1,500	1.0%
7	4,350	$ 107,760	$ 15,000	13.9%	$ 6,000	5.6%	$ 1,500	1.4%
8	3,975	$ 100,370	$ 15,000	14.9%	$ 6,000	6.0%	$ 1,500	1.5%
9	4,050	$ 108,440	$ 15,000	13.8%	$ 6,000	5.5%	$ 1,500	1.4%
10	4,300	$ 118,250	$ 15,000	12.7%	$ 6,000	5.1%	$ 1,500	1.3%
11	4,620	$ 129,360	$ 15,000	11.6%	$ 6,000	4.6%	$ 1,500	1.2%
12	4,750	$ 133,000	$ 15,000	11.3%	$ 6,000	4.5%	$ 1,500	1.1%
Total	53,650	$ 1,444,795	$ 180,000		$ 72,000		$ 18,000	
Lowest		97,600		10.0%		4.0%		1.0%
Highest		150,230		15.4%		6.1%		1.5%

Let us take the rent expense for example. It is set at $15,000 per month. When the revenues are at the highest in the sixth month with

$150,230, the rent was only 10% of the revenues. In the first month when the revenues were at its lowest with $97,600, the rent amount was still $15,000 but it was 15.4% of the revenues in that month. Similar trends can be easily identified in Table 5.5 with the depreciation expenses (Dep.) and in the interest expenses (Interest.). With these costs, the larger amount of revenues generated, the smaller amount of the cost will be incurred per unit sales and the operating profits will become larger. This point of view is very important in the analysis of the operating leverage, which will be explained in the next chapter.

5.7 MIXED COSTS

Up to this point, the nature of VCs and FCs has been introduced with a few examples from R&B Grill. If these were the only costs, it would be easy to determine R&B Grill's cost structure. It spends 33.5% of its revenues for the cost of sales, and additionally, it spends $22,500 for the fixed costs in total per month for the rent, depreciation, and interest. This finding can be presented as follows:

Total Costs = Variable Costs (33.5% of the revenues) + Fixed Costs ($22,500)

However, the Income Statement shows many other expenses beyond the VC and the FC. These additional cost items cannot be easily grouped into the VC or into the FC. They are the mix of the two containing unknown amount of fixed cost and the rest are made of VC. When a cost item is made of both VC and FC, it is called as MC. As introduced earlier, Payroll expense is a very good example. Many other costs listed in the Income Statements of any restaurant companies are of this nature and R&B Grill' is no exception.

As explained in the following section, restaurateurs can determine the amount of the FC hidden in individual MCs using the nature of the FC and the VC as introduced. By separating the amount of the FC and the VC in each MC, restaurateurs can develop the general cost structure of their business. Using the cost structure, they can project the future conditions identifying the amount of fixed cost involved in individual MCs, and the amount of VC that will be incurred. In restaurant businesses, one customer is the most widely used unit in this application. Let us use the monthly payroll expenses of R&B Grill to explain this further. The following Table 5.6 presents the list of the mixed costs.

Table 5-06(A) R&B Grill's Mixed Costs

Month	Customer Count	Revenues	Labor	Restaurant Operating	Marketing	Utility	Admin	R&M
1	3,750	$ 97,600	$ 31,000	$ 5,550	$ 3,400	$ 2,750	$ 6,420	$ 1,250
2	4,200	$ 115,600	$ 32,800	$ 6,130	$ 2,600	$ 2,850	$ 6,565	$ 1,430
3	4,450	$ 122,375	$ 32,200	$ 5,470	$ 1,800	$ 2,600	$ 6,310	$ 1,110
4	4,850	$ 133,475	$ 32,550	$ 5,625	$ 1,550	$ 2,450	$ 6,675	$ 1,175
5	4,980	$ 128,335	$ 34,600	$ 5,685	$ 1,400	$ 2,700	$ 6,450	$ 1,260
6	5,375	$ 150,230	$ 36,750	$ 8,425	$ 1,900	$ 3,700	$ 7,425	$ 2,075
7	4,350	$ 107,760	$ 31,800	$ 4,245	$ 1,170	$ 2,700	$ 6,185	$ 920
8	3,975	$ 100,370	$ 30,850	$ 3,295	$ 650	$ 2,500	$ 5,875	$ 850
9	4,050	$ 108,440	$ 32,750	$ 3,750	$ 850	$ 2,700	$ 6,350	$ 950
10	4,300	$ 118,250	$ 34,000	$ 4,270	$ 580	$ 2,750	$ 6,600	$ 800
11	4,620	$ 129,360	$ 36,000	$ 4,800	$ 450	$ 2,900	$ 7,100	$ 1,250
12	4,750	$ 133,000	$ 36,500	$ 6,050	$ 420	$ 3,150	$ 7,600	$ 1,750
Total	53,650	$ 1,444,795	$ 401,800	$ 63,295	$ 16,770	$ 33,750	$ 79,555	$ 14,820
Lowest	3,750	$ 97,600	$ 30,850	$ 3,295	$ 420	$ 2,450	$ 5,875	$ 800
Highest	5,375	$ 150,230	$ 36,750	$ 8,425	$ 3,400	$ 3,700	$ 7,600	$ 2,075

The expense titles presented in the table are group titles. Each includes multiple expense items as explained below:

- Labor expenses include salaries, wages, and other necessary employee benefits.
- Restaurant operating expenses include spending for vehicles, cleaning, equipment rental, smallware, pest control, printing, entertainment, decoration, and the like.
- Marketing expenses include promotion-related expenses.
- Utility expenses include electricity, gas, water and sewage, trash removal, and the like.
- Admin includes general and administrative expenses for back of the house functions, computer system–related spending, licenses and permits, telephone and communications, travel, professional services.
- Repair and maintenance (R&M) includes repair and maintenance-related expenses.

5.8 COST STRUCTURE OF INDIVIDUAL MIXED COSTS

The monthly payroll amount ranges from \$30,850 to \$36,750. When monthly expense amounts are compared with the customer counts of the month, a visible pattern emerges. When the customer count increases, the payroll amount also increases. As the payroll expense is considered as MC, it can be concluded that in each month the same amount of FC is spent and a certain amount of VC is additionally spent when one customer is served. This conclusion can be presented as shown in the following equation.

Monthly Payroll Expense = FC of the payroll + VC per customer × the # of the total customer

The equation above is usually presented in an algebraic form of $Y = a + bX$, where "Y" stands for the Expense amount; "a" indicates the amount of fixed cost involved; "b" indicates the average variable costs spent for each unit sales; and finally, the "X" indicates the total number of the unit sales, which is the customer count in this example. Although most textbooks on Managerial Accounting introduce the manual technique of the "High-Low Two point method," this book will bypass the manual technique and will introduce the application of the Microsoft Excel spreadsheet for this instead as presented in Table 5.6(B).

Table 5-06(B) R&B Grill's Mixed Costs Breakdown

Month	Customer	Pyrl	Oper	Mktg	Energy	Admin	R&M
1	3,750	$ 31,000	$ 5,550	$ 3,400	$ 2,750	$ 7,920	$ 1,250
2	4,200	$ 32,800	$ 6,130	$ 2,600	$ 2,850	$ 8,065	$ 1,430
3	4,450	$ 32,200	$ 5,470	$ 1,800	$ 2,600	$ 7,810	$ 1,110
4	4,850	$ 32,550	$ 5,625	$ 1,550	$ 2,450	$ 8,175	$ 1,175
5	4,980	$ 34,600	$ 5,685	$ 1,400	$ 2,700	$ 7,950	$ 1,260
6	5,375	$ 36,750	$ 8,425	$ 1,900	$ 3,700	$ 8,925	$ 2,075
7	4,350	$ 31,800	$ 4,245	$ 1,170	$ 2,700	$ 7,685	$ 920
8	3,975	$ 30,850	$ 3,295	$ 650	$ 2,500	$ 7,375	$ 850
9	4,050	$ 32,750	$ 3,750	$ 850	$ 2,700	$ 7,850	$ 950
10	4,300	$ 34,000	$ 4,270	$ 580	$ 2,750	$ 8,100	$ 800
11	4,620	$ 36,000	$ 4,800	$ 450	$ 2,900	$ 8,600	$ 1,250
12	4,750	$ 36,500	$ 6,050	$ 420	$ 3,150	$ 9,100	$ 1,750
Total	53,650	$ 401,800	$ 63,295	$ 16,770	$ 33,750	$ 97,555	$ 14,820
Lowest	3,750	$ 30,850	$ 3,295	$ 420	$ 2,450	$ 7,375	$ 800
Highest	5,375	$ 36,750	$ 8,425	$ 3,400	$ 3,700	$ 9,100	$ 2,075
FC (intercept)		$ 18,241	$ (3,778)	$ 3,107	$ 971	$ 4,915	$ (1,127)
VC/cust (slope)		$ 3.41	$ 2.02	$ (0.38)	$ 0.41	$ 0.72	$ 0.53
Correlation		0.77	0.70	(0.19)	0.57	0.66	0.66

Table 5.6(B) shows extended information on the bottom. Underneath the lowest and the highest values of each MC, it has additional three rows of new information. The first (additional) row says "FC (intercept)," followed by "VC/customer (slope)." Finally on the bottom row is "correlation." These are statistical parameters obtained using the "Regression Analysis" function of a spreadsheet program (Microsoft Excel has been used for this application). These parameters can be applied to business cost analysis. The formulas of Excel for this task are introduced later in this section.

The FC is of the same nature of the intercept ("a") in the equation of "$Y = a + bX$" in algebra. This is sometimes called the "constant" because of its unchanging nature. Because FCs remain the same, it is represented by "*a*" in the equation. The Total Cost of "*Y*" starts from the amount of FC ("*a*") when there are no revenues, thus no VC. When there is no VC, the bX becomes zero (0) and the *Y* becomes the same as the "*a*" which is the amount of the FC. The VC, represented by "*bX*," is further explained below.

The value of the "*b*" in the equation remains the same; but the entire value of "bX," which is the multiple of the "*b*" and "*X*," changes when the value of the "*X*" varies. Here, the "*b*" stands for the VC per one customer (thus, "VC/customer) and the "*X*" stands for the number of customers. When no customer is served, *X* becomes zero (0). Then, the VC is also zero (0). As pointed out, the business still has to spend its FC (presented as "*a*"). When one customer is served, the VC increases by the amount of "b," which is the average VC incurred when serving one customer. The amounts of the FC (a) and the VC per customer presented in Table 5.6(B) have been obtained by applying the built-in formula of the Excel program.

Using the built-in formula in Excel, these parameters can easily be determined without taking the calculation process manually. It is important, however, to understand what these parameters truly represent. The formula of determining the FC (or "intercept") of the Payroll Expense is "=intercept(the range of the Payroll Expenses (PYRL), the range of the Customer Count)" with no space in between. The Excel formula for the VC/customer (or "slope") is "=slope(the range of the PYRL, the range of the Customer Count)" again with no space in between. Once these formulae are correctly entered, the spreadsheet will calculate the necessary parameters and provide them instantly. They are, in the case of the Payroll, $18,241 for the FC and $3.41 for the VC per customer. The interpretation of this result is as follows:

"R&B Grill's monthly payroll expenses are made of $18,241 of the fixed amount, and additional amount of $3.41 for each customer served." In an algebraic format, it can also be presented as follows:

Total monthly Payroll Expenses = FC of $18,241 + $3.41 × the # of the customer during the month

5.9 HOW TO HANDLE THE VARIANCE BETWEEN THE THEORY AND PRACTICE

When this generalized structure is applied to the actual customer count of each month, the actual result will be slightly different; but the discrepancy will be small enough to accept. This closeness is presented on the bottom row of the table as "Correlation Coefficient." The Excel formula is "=correl(the range of one data set array, the range of the other data set array)" with no space in between of course. The result of 0.77 as presented on the table indicates that the closeness of the two data sets – Payroll expenses and the Customer count – are correlated with each other by around 77%. Other MCs can be divided into their FC and the VC per customers by applying the same formula.

One important fact to consider in this approach is to use common sense. The Marketing Expense shows the FC of $3,107 with the negative VC per customer value of –$0.38. The trend of the marketing expense is not similar to that of the Payroll Expenses or Operating Expenses. The direct prediction using this result would be "The maximum marketing expense of one month will be $3,107. From that amount, it will go down by $0.38 per additional customer served." One possible explanation for this is that marketing expenses are usually spent when the business gets slow and after spending expenses in marketing and promoting activities, business will get busier. For this reason, the amount spent on this expense is not highly correlated with the customer count of the relevant month. This explains the low and negative correlation coefficient at –0.19 indicates the low level of interdependence of the two variables. To project an expense of this nature, the management must use its own discretion. Many operators tend to allocate a fixed amount into this type of expenses in their budgeting process.

Other items that require caution are "Restaurant Operating Expenses" and "Repair and Maintenance." They both show negative FC. This is just the result of the calculation of the given data. There are no such things as negative FC. When the result turns out this way, the management also must apply its own common sense to set up their plan.

5.10 BREAK-EVEN ANALYSIS

In this chapter, different types of costs and their unique nature are introduced. Along with them, a convenient method has been introduced to break down individual costs into the FC amount and the VC. The result of these approaches has been summarized into a simple algebraic equation that says a tMC consists of a FC plus the multiple of the VC per unit (customer, in the example) and the total number of the unit served.

In fact, this approach can be applied to the entire costs to determine the overall cost structure of the business and further to identify the break-even point of sales. With the projected break-even sales under the existing cost structure, they can prepare more specific plans to achieve better results by adjusting their cost structure or pricing policies.

The next is to conduct the break-even analysis of R&B Grill with the same data. This time, apply the same Excel formulae of the intercept for the FC and of the slope for the VC per customer. Table 5.7 has two additional columns at the end. One is the Total Costs (TL costs) and the next one is the profit or loss of each year (P/L).

Referring to the bottom of the "Total Cost" column, the application of the Excel formula has calculated the total FC to be $52,619. The unit VC per customer is $14. This result means that R&B Grill has to spend $52,619 of FCs regardless of its revenues and to that amount $14 of additional VC will be added when one customer is served. By comparing the total revenues of $1,444,195 with the annual customer count, the average check price is $26,91.

To summarize, R&B Grill generates $26.91 and spends $14 directly for each customer. When one customer is served, the company will have about $13 ($12.91 to be exact) of profit, which is called the contribution margin per customer, is generated. To break-even, this amount ($12.91 of contribution margin per customer) must be accumulated to cover all other FCs that are $52,619 in total. Therefore, dividing the amount of the FCs in total ($52,619) by the $12.91 of contribution margin per customer, the necessary number of customers to serve is 4,076. Then, the total revenues will be approximately $109,685 (average check price of $26.91 x 4,076 customers). Out of this amount, the total VCs of $57,064 ($14 VC per customer × 4,076 customers) will be spent and the business will be left with $52,621 for additional spending. This is will be enough to cover the total FCs of $52,619. A small amount of changes is left in this calculation; but it is acceptable to call this a break-even point sales amount.

Table 5-07 R&B Grill's Annual Operations Breakdown – Total Costs & EBT

Mo.	Custo-mer	Revenues	COS	Labor	RO	Mktg	Utility	Admin	R&M	Rent	Dep.	Int.	TL costs	EBT
1	3,750	97,600	36,500	31,000	5,550	3,400	2,750	7,920	1,250	15,000	6,000	1,500	110,870	(13,270)
2	4,200	115,600	40,300	32,800	6,130	2,600	2,850	8,065	1,430	15,000	6,000	1,500	116,675	(1,075)
3	4,450	122,375	39,500	32,200	5,470	1,800	2,600	7,810	1,110	15,000	6,000	1,500	112,990	9,385
4	4,850	133,475	42,500	32,550	5,625	1,550	2,450	8,175	1,175	15,000	6,000	1,500	116,525	16,950
5	4,980	128,335	43,500	34,600	5,685	1,400	2,700	7,950	1,260	15,000	6,000	1,500	119,595	8,740
6	5,375	150,230	46,500	36,750	8,425	1,900	3,700	8,925	2,075	15,000	6,000	1,500	130,775	19,455
7	4,350	107,160	40,500	31,800	4,245	1,170	2,700	7,685	920	15,000	6,000	1,500	111,520	(4,360)
8	3,975	100,370	34,000	30,850	3,295	650	2,500	7,375	850	15,000	6,000	1,500	102,020	(1,650)
9	4,050	108,440	36,500	32,750	3,750	850	2,700	7,850	950	15,000	6,000	1,500	107,850	590
10	4,300	118,250	38,000	34,000	4,270	580	2,750	8,100	800	15,000	6,000	1,500	111,000	7,250
11	4,620	129,360	42,000	36,000	4,800	450	2,900	8,600	1,250	15,000	6,000	1,500	118,500	10,860
12	4,750	133,000	44,500	36,500	6,050	420	3,150	9,100	1,750	15,000	6,000	1,500	123,970	9,030
Total	53,650	1,444,195	484,300	401,800	63,295	16,770	33,750	97,555	14,820	180,000	72,000	18,000	1,382,290	61,905
FC				18,241	(3,778)	3,107	971	4,915	(1,127)				52,619	
VC				3.41	2.02	(0.38)	0.41	0.72	0.53				14.00	

Correlation Coefficient – 0.86.

RO – Restaurant operating; Mo. – Month.

This approach is often presented in an algebraic format in most books. The following describes the logic presented above in a step-by-step mathematical explanation.

1. First, the revenues can be presented as the multiple of the average check price ($26.91) and the customer count (X). Using the alphabet variables X for the customer count, the revenues amount can be presented as $26.91 × X.

2. The cost structure has been identified for the total FCs at $52,619 and the VCs at $14 per customer. The total VCs can be presented as $14 × Customer Count ($X$). Using the result of this finding, the total costs (FC + VC) can be presented as the total FC ($52,619) plus the total VCs [$14 times (×) Customer Count (X)]. In an equation form, this will be presented as: Total Costs = $52,619 + $14 × X.

3. At the break-even point, the sales amount ($P \times X$) must be the same as the Total Costs ($52,619 + $14 × X). This is, in equation form, presented as: $26.91 × X = $52,619 + $14 × X.

4. Both sides of the equation have the same variable X, which can be moved to the same side as follows: ($26.93 − $14) × X = $52,619. When the equation is solved for X, it is 4,076 customers.

It is important to understand the meaning of the difference between the Average Check of $26.91 and the total VCs of $14. It is the necessary amount of profit (contribution margin) that should be accumulated to the amount of FC to break-even. The conclusion of this is usually presented in the formula of calculating the necessary customer count to break-even as: The Fixed Cost amount/Contribution Margin.

CHAPTER 6

FORECASTING AND PLANNING

6.1 INTRODUCTION

This chapter introduces forecasting and planning of a restaurant business. Both are future-oriented activities necessary to navigate the business to achieve the ultimate goal of providing positive return to the investors. In this context, both can be summarized as setting the ultimate goal and designing necessary activities to achieve it.

More specifically, forecasting involves setting up expected goals such as the amount of revenues to generate and the amount of expenses that will be incurred. Planning, on the other hand, involves practical blueprints to guide the operational activities, such as labor scheduling, purchasing necessary inventory and supplies, and replenishing necessary amount of cash to name just a few. These must be smoothly orchestrated to achieve the goals set by the forecast. No matter how thoroughly plans are prepared, there will be discrepancies in actual results. Managers of a restaurant business must analyze the discrepancy and adjust the plans accordingly.

The most frequently introduced forecasting techniques include the naïve method, moving average method, causal forecasting method, and exponential smoothing method. With their own advantages and disadvantages, these techniques are widely used by the industry practitioners. The following is a brief introduction of these methods.

6.2 NAÏVE METHOD

This is the simplest method of forecasting sales or profits by applying the opinion of the practitioner. When a restaurateur says "My goal is to achieve 10% growth in sales next year," the restaurateur is using a naïve method of forecasting. Simple and easy to apply, this method can be used for short-term goal when business conditions are obvious to forecast. Our example of forecasting the revenues of R&B Grill in Y-2 uses this method because it is considered easy to achieve the 15% increase in the customer count, and 10% increase in the average check price during the second year of business from those of the first year (Refer to Table 6.1 "Customer Count Projection"). This is an example of how to apply the naïve method of forecasting. When a business matures, however, this method may not be effective. Practitioners must apply this method with caution.

6.3 MOVING AVERAGE METHOD

This method assumes the average of revenues and expenses of a fixed number of past months or years will continue in the following month or year. Many research organizations use this method to project the future occupancy rates of the hotel industry. It is common that the average of 48 to 60 or even 72 monthly occupancy rates is used to project the following month's occupancy rates in a region. The same approach is also used to forecast the Average Daily Rates. Projecting the number of tourists in a certain region also often adopts this method. The limitation of this method, however, is that the long-term forecasting will follow the previous trend. If the previous trends were increasing, the future will also continue to increase until it stops growing. If, on the other hand, the previous trends were declining, the future will also decline. For this reason, this method may be used for short-term projection. It may not be reliable for long-term forecasting.

When a business has a large database that covers long-term, it may adopt this method to forecast specific months or even weeks to project the result by identifying the pattern of the same months or weeks during the previous years.

6.4 CAUSAL FORECASTING METHOD

This method, designed to identify a pattern between causal factors that affect the result, requires reasonable interpretation of business conditions that are believed to affect the result. For example, weather conditions like temperature or humidity are considered as important factors that determine the sales of certain items such as cold drink or hot items by influencing the customers' desire to purchase. In this example, the number of customers is considered as the result (or dependent variable *Y*) and the temperature or humidity is considered as the cause (or independent variable *X*).

Historical records of these variables are applied to determine the pattern by identifying the value of "*a*" and "*b*", which is presented in an equation ($Y = a + b \times X$) to project the future. The interpretation of this equation is "the result ("*Y*" which is Customer Count, for example) starts from "*a*", and grows by "*b*" when the temperature is "*X*." This method was introduced in the previous chapter when mixed costs were divided into the fixed cost (FC) and the variable cost (VC) per customers to project the future mixed cost spending. At that time, each cost amount was considered as the result, which was influenced by the number of customers of each month. To apply this method, it is recommended to use a spreadsheet program (such as MS Excel or other statistical software).

Later in this chapter, this method will be adopted to forecast the second year (Y-2) of R&B Grill for its revenues and individual expenses. The causal factor used in this approach is the expected number of customers of each month.

6.5 EXPONENTIAL SMOOTHING METHOD

This method adjusts the discrepancy between the historical forecast and the actual result by applying an index called as the "smoothing constant." Periodical analyses of the discrepancy are required in this approach along with the application of spreadsheet programs to determine the value of the smoothing constant. The technical details of this approach will not be covered in this book due to the complexity involved in obtaining the smoothing constant.

Each technique introduced above has its own strengths and weaknesses and all of them are using historical data under the assumption that past conditions will continue in the future. None of these methods, however, can be a sole magic stick in business forecasting by itself. Practitioners must be able to take the advantage of the strengths of each technique and apply them to their unique situation as a useful guide instead of worshiping them as an absolute truth.

This chapter presents practical examples of forecasting by applying the naïve method and the causal method together using the material introduced in previous chapters. It starts with the revenues forecast.

6.6 REVENUES FORECAST

In restaurant operations, revenues are directly related to the number of customers served and the maximum number of customers that can be served is determined by the size of the property. For this reason, forecasting revenues must be preceded by establishing a detailed operating schedule for each day of the week and the hours of operations of each day. Based on this, the expected number of customers must be projected and the menu items ordered by one customer must be applied. Of course, the menu items to be served must be determined with accurate standard recipes and the ideal cost (as introduced in Chapter 5). With the projected customer traffic of each day, the necessary labor schedule can also be prepared. Once these steps completed, the annual revenues will be forecast, along with the ideal food costs and the labor costs. The detailed steps of these activities are explained in the following sequence with R&B Grill which has 100 seats.

6.6.1 Daily Operating Hours with Expected Customer Count, Revenues, and Food Cost

The next task is to determine the operating hours of each day. It is a common practice for a restaurant to open longer hours during weekends. For this reason, it is a good idea to develop a weekly schedule and expand it to the entire year. The following is a scenario of R&B Grill's weekly operations.

Monday through Thursday:	from 4:00 p.m. until 10:00 p.m. (6 hours)
Fridays and Saturdays:	from 4:00 p.m. until 11:00 p.m. (7 hours)
Sundays:	from 3:00 p.m. until 10:00 p.m. (7 hours)

According to the scenario, R&B Grill runs 45 hours each week. In practice, it is desirable to use historical data from point of sales (POS) system with the expertise of seasoned operators in determining the most reasonable number of customers of each time zone of every day when developing detailed forecasting and subsequent planning. This projection needs to be conservatively prepared.

This section will introduce a simplified scenario for ease and convenience. According to the operating schedule, R&B Grill only serves dinner from 4:00 p.m. during the week, and from 3:00 p.m. on Sundays. Customers can be accommodated until the closing time.

The restaurant has 100 seats, and it operates every day. Table 6.1 is the original forecasting prepared at the time of opening. It shows the expected number of customers each day. The last three rows present the daily total customer counts, total seats, and the seat turnover of each day. According to this scenario, during the weekdays its seat turnover is as low as 0.71 to 1.14. The busiest day is Saturday with the seat turnover of 2.82. In total, the weekly customer count is 980, and the average weekly seat turnover is 9.8 as presented in the "Total" column.

In the following table are presented a few scenarios based on naïve forecasting method. As the first line (a) indicates, if the customer counts increase by 15% in the next year, it will serve 1,127 customers per week and finally 58,604 customers in 1 year (52 weeks). The next line (b) is also a naïve forecast that expects 10% increase in the average check price. If this projection works out as forecast, the average check price during the next year will be $29.60. If both projections work out as forecast, the total revenues will be $1,734,737 that is over 20% increase from the first year (Y-1). For a new business that is expected to grow rather rapidly, the naïve method may be applied with ease. However, once the business grows into its mature stage, the moving average may be more appropriate.

Table 6-01 Customer Count and Sales Projection

Hours	Mon	Tue	Wed	Thu	Fri	Sat	Sun	Total
3:00–4:00							5	5
4:00–5:00	15	10	10	12	15	20	8	90
5:00–6:00	15	12	16	15	24	26	18	126
6:00–7:00	22	25	25	28	36	48	16	200
7:00–8:00	24	24	22	25	48	60	12	215
8:00–9:00	22	20	20	18	36	60	8	184
9:00–10:00	10	8	15	16	24	48	4	125
10:00–11:00					15	20		35
Total Customers	108	99	108	114	198	282	71	980
Available seats	100	100	100	100	100	100	100	
Seat Turnover	1.08	0.99	1.08	1.14	1.98	2.82	0.71	9.8

All projections below are rounded numbers.

a: Forecast of 15% increase of weeky customer count (980 x 1.15)	1,127
Annual customer count (52 weeks)	58,604
b: Forecast per capita spending with 10% increase ($26.91 x 1.10)	$ 29.60
c: Annual revenues – 52 weeks (58,064 x $29.60)	$ 1,734,737
d: Revenues growth rate from the Y-1 ($1,444,195)	20.1%

6.7 FORECASTING THE IDEAL FOOD COST

Along with the revenues projection as presented here, food cost can also be projected. Since the food cost is a VC that must be close to the ideal food cost developed in the Chapter 5 (refer to Table 5.2 Sale Mix and Ideal Food Cost Percentage), R&B Grill's food cost must be close to 32.4% with reasonably acceptable discrepancy. We will use 33.5% as the target food cost percentage to be conservative. Operators must pay close attention to their food costs and make sure their inventory controls are effectively conducted to maintain the cost percentage at this level.

6.8 FORECASTING LABOR COSTS

The next item to forecast is the labor costs. Many restaurateurs argue that the labor cost in a restaurant business should be maintained at around one-third or lower of the total revenues. Although this rule of thumb provides an easy threshold in controlling the labor costs, it can be projected easier when compared with the customer count rather than just the amount of total revenues. This approach will give operators better information regarding how to accurately address their labor productivity. To complete this, the management must have a fully developed organization chart with clearly defined task list and job description for each position. Then, labor schedules must be prepared according to the operational plan.

The customer count information presented in Table 6.1 must be used as the base of the employee schedule for the front of the house and the back of the house as well. To this, salaries and other labor-related expenses must be added to complete the labor cost forecast. Of course, the result of this must be compared with the projected revenues to be considered acceptable.

As mentioned earlier, the labor cost in a restaurant is a mixed cost. Usually managers and supervisory personnel are paid salaries at fixed amounts, while many others are paid by the hour. However, in restaurant operations, once a labor schedule is established, it will be hard to change it when actual business situation deviates from the forecast. If scheduled employee hours are reduced often due to unexpected slow conditions, it will be hard to keep good employees. Hence, forecasting and planning must be carefully conducted.

6.8.1 Wage Forecasting of the Front of the House

Forecasting wage amount must be based on the customer count as projected in Table 6.1. Once customer counts are projected reasonably by the hour of operation, the management must schedule enough employees to cover the necessary workload. Scheduling must follow the internal policy with the productivity and the quality of service in mind. For example, if the internal policy is to assign one server to three tables or 9.12 customers, whichever is larger, the management must schedule enough servers to cover all customers with their full attention.

The wage forecasting of the front of the house staff is presented in Table 6.2 (Wage Projection – Front of The House). The table is made of seven daily projection sub-tables (Table 6.2(A)) and the consolidated weekly wage total table (Table 6.2(B)). The seven daily sub-tables can be used for the operators, but the consolidated weekly table can be used for accounting and analysis purposes.

Table 6-02(A) Wage Projection – Front of the House

Monday Projection					
Hours	Cust.*	In hall**	Server hours***		Total Hours
3:00–4:00			2		2
4:00–5:00	7	12	2		2
5:00–6:00	10	18	2	1	3
6:00–7:00	17	28	2	2	4
7:00–8:00	22	32	2	2	4
8:00–9:00	20	25	2	2	4
9:00–10:00	10	10	1	2	3
10:00–11:00				2	2
Total	86				
# of hours			13	11	24
Wage per hour – tipped employee					$ 5.50
Total daily wage					$132.00

Friday Projection					
Hours	Cust.*	In hall**	Server hours***		Total Hours
3:00–4:00			3		3
4:00–5:00	15	27	3		3
5:00–6:00	24	42	3	3	6
6:00–7:00	36	60	3	3	6
7:00–8:00	48	66	3	3	6
8:00–9:00	36	48	3	3	6
9:00–10:00	24	31	2	2	4
10:00–11:00	15	15	2	2	4
11:00–12:00				2	2
Total	198				
# of hours			22	18	40
Wage per hour					$ 5.50
Total daily wage					$220.00

Tuesday Projection					
Hours	Cust.*	In hall**	Server hours***		Total Hours
3:00–4:00			2		2
4:00–5:00	10	16	2		2
5:00–6:00	12	24	2	2	4
6:00–7:00	25	37	2	3	5
7:00–8:00	24	34	2	3	5
8:00–9:00	20	24	2	2	4
9:00–10:00	8	8		2	2
10:00–11:00				2	2
Total	99				
# of hours			12	14	26
Wage per hour					$ 5.50
Total daily wage					$143.00

Saturday Projection					
Hours	Cust.*	In hall**	Server hours***		Total Hours
3:00–4:00			4		4
4:00–5:00	20	33	4		4
5:00–6:00	26	50	4	3	7
6:00–7:00	48	78	4	3	7
7:00–8:00	60	90	4	3	7
8:00–9:00	60	84	4	3	7
9:00–10:00	48	58	4	3	7
10:00–11:00	20	20		3	3
11:00–12:00				3	3
Total	282				
# of hours			28	21	49
Wage per hour					$ 5.50
Total daily wage					$269.50

Table 6-02(A) (Continued)

Wednesday Projection					
Hours	Cust.*	In hall**	Server hours***		Total Hours
3:00–4:00			2		2
4:00–5:00	10	18	2		2
5:00–6:00	16	28	2	3	5
6:00–7:00	25	36	2	3	5
7:00–8:00	22	32	2	3	5
8:00–9:00	20	27	2	3	5
9:00–10:00	15	15		3	3
10:00–11:00				2	2
Total	108				
# of hours			12	17	29
Wage per hour					$ 5.50
Total daily wage					$159.50

Sunday Projection					
Hours	Cust.*	In hall**	Server hours***		Total Hours
2:00 – 3:00			1		1
3:00 – 4:00	5	9	1		1
4:00 – 5:00	8	17	1	2	3
5:00 – 6:00	18	26	1	2	3
6:00 – 7:00	16	22	1	2	3
7:00 – 8:00	12	16	1	2	3
8:00 – 9:00	8	10		2	2
9:00 – 10:00	4	4		2	2
10:00 – 11:00				2	2
Total	66				
# of hours			6	14	20
Wage per hour					$ 5.50
Total daily wage					$110.00

Thursday Projection

Hours	Cust.*	In hall**	Server hours***		Total Hours
3:00–4:00			2		2
4:00–5:00	12	19	2		2
5:00–6:00	15	29	2	2	4
6:00–7:00	28	40	2	2	4
7:00–8:00	25	34	2	2	4
8:00–9:00	18	26	2	2	4
9:00–10:00	16	16		2	2
10:00–11:00				2	2
Total	114				
# of hours			12	12	24
Wage per hour					$ 5.50
Total daily wage					$132.00

* The customer counts are the ones that leave by the end of each hour.

** In-hall customer count of each hour is estimated with a half of the next hour customer added to the current ones. For example, during the hour of 6-7, 17 customers are finishing their meal, and 11 additional customers are being served.

*** Server hours stand for the number of servers working in each hour. For example, from 5 to 6 o'clock, 4 servers are scheduled to work to serve 28 customers in the dining hall. During this one hour, 4 hours' payroll is being incurred.

More details are introduced in the text.

The daily sub-tables show each day's labor schedule. The first column shows the breakdown of the operating hours of the day. The second column indicates the anticipated number of customers arriving during each hour, which has been presented in Table 6.1. The third column "in hall" represents the potential number of customers that need to be served. It is expected that only a small number of customers will come in for dinner at earlier hours each day. As time goes by, more customers will arrive and eventually the number of arrival will reduce. About a half of those who came in each hour are expected to stay over to the next hour. Hence, the third column "in hall" shows a larger number of the customer that needs service. This is based on an arbitrary estimate that customers spend about one hour on average. Operators need to adjust this estimate depending on their own situations.

In the dining room, two servers are scheduled to arrive one hour before the opening to prepare and serve when early birds arrive. By 5:00 p.m., one or two additional servers are scheduled to arrive, so that three to four employees in total will serve guests during the peak time. By 9:00 p.m. when the operations slow down, the early shift servers will leave, and the late shift will continue their work until closing. Since a server is a tipped position, the hourly wage paid can be lowered. In this projection, it is set at $5.50 to be conservative to accommodate the current issues related to the minimum wage. In this structure, Monday's total amount of wage for the front of the house will be $132 that covers 24 employee hours for the front of the house function.

Table 6-02(B) Weekly Wage Total – Front of the House

Day	Hour	Wage	Amount
Monday	24	$ 5.50	$ 132
Tuesday	26	$ 5.50	$ 143
Wednesday	29	$ 5.50	$ 160
Thursday	24	$ 5.50	$ 132
Friday	40	$ 5.50	$ 220
Saturday	49	$ 5.50	$ 270
Sunday	20	$ 5.50	$ 110
Weekly Total	212		$ 1,166
Annual Total (52 wks)			$ 60,632

* These are tipped employees.

All other days' schedules follow the same format. Since Fridays and Saturdays have longer hours of operations, scheduling must be extended. Based on this scenario, the total annual amount of wages for the front of the house is $60,346 as presented in Table 6.2(B) (Weekly Wage Total – Front of the House).

6.8.2 Wage Forecasting of the Back of the House

The technique of scheduling labor in the back of the house is identical to that of the front of the house. The only difference is that the hours of working are different due to the nature of the work. The back of the house of a restaurant conducts preparatory work. For this reason, back of the house employees must start much earlier than the front as shown in Table 6.3(A). Like the front of the house, there are two shifts. One shift starts and ends earlier, while the other shift does late. As these positions do not receive tips, at least legal minimum wage must be applied in scheduling. To accommodate recent developments of the labor-related legal issues, the hourly wage of R&B Grill is estimated at $14 per hour for this scenario. Accordingly, this restaurant will incur $560 for the back of the house wage on Mondays by providing 40 employee hours. This way, the annual wage amount is $144,872 as shown in Table 6.3(B).

6.8.3 Salary Forecasting

Managers and supervisory personnel are usually paid salaries per year. These personnel are responsible for general tasks to run the entire operation beyond dealing with customers. In our example, three salaried positions – the general manager, one floor manager, and one kitchen manager – will be maintained at $150,000 combined altogether. To this, emergency reserve funds of $50,000 are set aside to prepare for unexpected occasions when additional labor is needed.

6.8.4 Payroll-Related Costs

One important cost that can be easily overlooked is the extra costs involved in labor. These extra costs include payroll taxes, overtime pay, insurance and other benefits, and miscellaneous expenses necessary to retain a proper labor force. In our example, $101,305, which is about 25% of the net labor cost, is allocated to cover these costs. Table 6.4 shows the summarized result of the labor cost based on the examples:

Table 6-03(A) Wage Projection – Back of the House

Monday Projection					
Hours	Cust.*	In hall**	Server hours***		Total Hours
12:00–1:00			3		3
1:00–2:00			3		3
2:00–3:00			3	2	5
3:00–4:00			3	2	5
4:00–5:00	15	22	3	2	5
5:00–6:00	15	26	3	2	5
6:00–7:00	22	34	3	2	5
7:00–8:00	24	35	3	2	5
8:00–9:00	22	27		2	2
9:00–10:00	10	10		2	2
10:00–11:00					
Total	108				
# of hours			24	16	40
Wage per hour					$ 14.00
Total daily wage					$ 560.00

Friday Projection					
Hours	Cust.*	In hall**	Server hours***		Total Hours
12:00–1:00			3		3
1:00–2:00			3		3
2:00–3:00			3	2	5
3:00–4:00	15		3	2	5
4:00–5:00	24	42	3	2	5
5:00–6:00	36	60	3	2	5
6:00–7:00	48	66	3	2	5
7:00–8:00	36	48	3	2	5
8:00–9:00	24	31		2	2
9:00–10:00	15	15		2	2
10:00–11:00					
Total	198				
# of hours			24	16	40
Wage per hour					$ 12.00
Total daily wage					$480.00

Tuesday Projection Hours	Cust.*	In hall**	Server hours***		Total Hours
12:00–1:00			3		3
1:00–2:00			3		3
2:00–3:00			3	2	5
3:00–4:00			3	2	5
4:00–5:00	10	16	3	2	5
5:00–6:00	12	24	3	2	5
6:00–7:00	25	37	3	2	5
7:00–8:00	24	34	3	2	5
8:00–9:00	20	24		2	2
9:00–10:00	8	8		2	2
10:00–11:00					
Total	99				
# of hours			24	16	40
Wage per hour					$ 12.00
Total daily wage					$ 480.00

Saturday Projection Hours	Cust.*	In hall**	Server hours***		Total Hours
12:00–1:00			3		3
1:00–2:00			3		3
2:00–3:00			3	2	5
3:00–4:00			3	2	5
4:00–5:00	20	33	3	2	5
5:00–6:00	26	50	3	2	5
6:00–7:00	48	78	3	2	5
7:00–8:00	60	90	3	2	5
8:00–9:00	60	84	3	2	5
9:00–10:00	48	58	3	2	5
10:00–11:00	20	20		2	2
Total	282				
# of hours			30	18	48
Wage per hour					$ 12.00
Total daily wage					$576.00

Table 6-03(A) (Continued)

Wednesday Projection Hours	Cust.*	In hall**	Server hours***		Total Hours
12:00–1:00			3		3
1:00–2:00			3		3
2:00–3:00			3	2	5
3:00–4:00			3	2	5
4:00–5:00	10	18	3	2	5
5:00–6:00	16	28	3	2	5
6:00–7:00	25	36	3	2	5
7:00–8:00	22	32	3	2	5
8:00–9:00	20	27		2	2
9:00–10:00	15	15		2	2
10:00–11:00					
Total	108				
# of hours			24	16	40
wage per hour					$ 12.00
Total daily wage					$ 480.00

Sunday Projection Hours	Cust.*	In hall**	Server hours***		Total Hours
12:00–1:00			2		2
1:00–2:00			2		2
2:00–3:00			2	2	4
3:00–4:00	5	9	2	2	4
4:00–5:00	8	17	2	2	4
5:00–6:00	18	26	2	2	4
6:00–7:00	16	22	2	2	4
7:00–8:00	12	16	2	2	4
8:00–9:00	8	10		2	2
9:00–10:00	4	4		2	2
10:00–11:00					
Total	71				
# of hours			16	16	32
wage per hour					$ 12.00
Total daily wage					$384.00

Thursday Projection

Hours	Cust.*	In hall**	Server hours***		Total Hours
12:00–1:00			3		3
1:00–2:00			3		3
2:00–3:00			3	2	5
3:00–4:00			3	2	5
4:00–5:00	12	19	3	2	5
5:00–6:00	15	29	3	2	5
6:00–7:00	28	40	3	2	5
7:00–8:00	25	34	3	2	5
8:00–9:00	18	26	3	2	5
9:00–10:00	16	16		2	2
10:00–11:00					
Total	114				
# of hours			27	16	43
wage per hour					$ 12.00
Total daily wage					$ 516.00

Table 6-03(B) Weekly Wage Total – Back

Day	Hour	Wage	Amount
Monday	40	$ 14.00	$ 560
Tuesday	40	$ 14.00	$ 560
Wednesday	40	$ 14.00	$ 560
Thursday	43	$ 14.00	$ 602
Friday	12	$ 14.00	$ 168
Saturday	12	$ 14.00	$ 168
Sunday	12	$ 14.00	$ 168
Weekly Total	199		$ 2,786
Annual Total (52 wks)			$ 144,872

Table 6-04 Total Labor Cost Breakdown

Salaries	$ 150,000
Wage – Front	$ 60,346
Wage – Back	$ 144,872
Contingency Labor	$ 50,000
Net Labor Costs	$ 405,218
Labor related	$ 101,605
Total Labor Costs	$ 506,823
Total Revenues Projected	$ 1,736,026
Labor cost %	29.2%

Table 6.4 shows that the projected total labor cost is $506,823. Compared with the budgeted revenues of $1,734,737 that was presented in Table 6.1, the ideal labor cost will be slightly under 30% (29.2% to be exact). With the ideal food cost estimated at 32.4%, the prime costs will be under 62% leaving the 38% of the revenues for other expenses. This is lower than the rule of thumb widely used in the industry. R&B Grill seems in the right direction.

6.9 FORECASTING ALL OTHER COSTS (EXPENSES) USING EXCEL

The prime costs, as presented so far, must be forecast with the expected revenues because they are directly related to the number of customers to serve. Other costs, although they are not directly related, still can be forecast based on the estimated customer count. In the previous chapter, the FC amount and the VC per customer included in each cost have been identified using the technique of the regression analysis as replicated in the Table 6.5. These results will be used to project the amount of the individual expense items in this section.

The two bottom rows in Table 6.5 contain the results of the regression analysis technique that have provided the FC amount and the VC per customer included in individual cost items. The cost of sales is considered as VC with no fixed amount of spending when no customers are served. Hence, it does not have FC and VC per customer. The last three cost items – rent, depreciation, and interest – do not have variable costs involved because they are FCs. All others are mixed costs that have both types of costs.

The regression technique is commonly used in the Causal Forecasting Method. It establishes the causal relationship between two variables assuming one variable causes the other to change as the result. In this example, the assumption established is the number of customers causes each mixed cost item to change. However, each cost has a fixed amount to be spent even when no customers are served. Under this assumption, for the payroll expense, for example, \$18,241 must be spent as a FC every month and \$3.41 must be additionally spent for every customer served. This is presented in an equation as follows.

Total Monthly Payroll Expense = Fixed Cost amount + VC/customer × expected # of Customers

= \$18,241 + \$3.41 × expected the # of customers

For example, if R&B Grill expects to serve 5,000 customers in the next month, it must be prepared to spend \$18,241 + \$3.41 x 5,000 = \$35,291 for the monthly payroll expense. To forecast the annual payroll expense amount, the monthly FC (\$18,241) needs to be multiplied by 12, and the number of customers expected in the entire year should be multiplied by the variable cost per customer (\$3.41). This simple application is possible because the regression analysis technique is used to identify two constant amounts – monthly FC amount and the VC per customer – from the historical trend. As long as the management of R&B Grill maintains

Table 6-05 The R&B Grill's Cost Breakdown with Regression Analysis Technique

Mo.	Customer	Revenues	COS	Pyrl	RO	Mktg	Energy	Admin	R&M	Rent	Dep.	Int.
1	3,750	97,600	36,500	31,000	5,550	3,400	2,750	7,920	1,250	15,000	6,000	1,500
2	4,200	115,600	40,300	32,800	6,130	2,600	2,850	8,065	1,430	15,000	6,000	1,500
3	4,450	122,375	39,500	32,200	5,470	1,800	2,600	7,810	1,110	15,000	6,000	1,500
4	4,850	133,475	42,500	32,550	5,625	1,550	2,450	8,175	1,175	15,000	6,000	1,500
5	4,980	128,335	43,500	34,600	5,685	1,400	2,700	7,950	1,260	15,000	6,000	1,500
6	5,375	150,230	46,500	36,750	8,425	1,900	3,700	8,925	2,075	15,000	6,000	1,500
7	4,350	107,760	40,500	31,800	4,245	1,170	2,700	7,685	920	15,000	6,000	1,500
8	3,975	100,370	34,000	30,850	3,295	650	2,500	7,375	850	15,000	6,000	1,500
9	4,050	108,440	36,500	32,750	3,750	850	2,700	7,850	950	15,000	6,000	1,500
10	4,300	118,250	38,000	34,000	4,270	580	2,750	8,100	800	15,000	6,000	1,500
11	4,620	129,360	42,000	36,000	4,800	450	2,900	8,600	1,250	15,000	6,000	1,500
12	4,750	133,000	44,500	36,500	6,050	420	3,150	9,100	1,750	15,000	6,000	1,500
Total	53,650	1,444,795	484,300	401,800	63,295	16,770	33,750	97,555	14,820	180,000	72,000	18,000
FC (Fixed Cost)				18,241	(3,778)	3,107	971	4,915	(1,127)	15,000	6,000	1,500
VC/customer				3.41	2.02	(0.38)	0.41	0.72	0.53	–	–	–

RO – Restaurant Operating; Mo. – Month.

the same labor productivity, this equation can be used to project the labor cost of each month.

In our customer count forecast using the naïve method in the beginning of this chapter, the expected annual customer count was estimated to be 58,604 (refer to Table 6.1). If this forecast works out, according to the equation obtained above, the total labor cost will be:

$$(\$18{,}241 \times 12 \text{ months}) + \$3.41 \times 58{,}604 = \$218{,}080$$

The result obtained above is much lower than the payroll costs forecast prepared according to the labor schedule. It was $506,823. The reason for this huge discrepancy was caused by the adjustments made in the pay rate to accommodate the impact of the minimum pay-related regulations. In addition to it, a couple of salaried positions were newly created. In addition, extra $50,000 has been set aside to handle any contingency situation related to the labor situation. This is an example of very conservative projection the management of a business must consider in their budgeting process.

6.10 OTHER COSTS AND PRO FORMA INCOME STATEMENT

All other expenses besides the prime costs also must be forecasted applying the same technique. If everything works out as forecast described, the result of the annual operations during the second year (Y-2) will be as presented in the following pro forma Income Statement (Table 6.6). It shows the list of all expenses in group names such as Restaurant Operating Expenses, Marketing Expenses, Energy (Utility) Expenses, Administrative Expenses, Repair and Maintenance Expenses, Rent, Depreciation, and finally Interest Expenses. It must be noticed that the titles of some expenses are slightly altered. The Insurance Expenses have been included into the Administrative Expenses. It is common practice in the restaurant industry to put multiple expense items into a representative group for simplicity in their presentation of the operational results.

Table 6-06 Pro Forma Income State of the Y-2

R&B Grill						
	Y-1 Actual		Y-2 Forecast			
	Y-1	V.A. %	FC	VC/cust	Y-2	V.A. %
			Customer #	58,604		
Revenues	14,444,195	100.0%	Avg. Check*	$ 29.60	1,734,678	100.0%
Cost of Sales	484,300	3.4%	Cost %	33.5%	581,117	33.5%
Gross Profit	13,959,895	96.6%			1,153,561	66.5%
Operating Expenses			FC (annual)	VC/cust		
Labor Expense	401,800	2.8%	per labor schedule**		506,823	29.2%
Restaurant Operating Expense	63,295	0.4%	***	$ 1.25	73,255	4.2%
Marketing Expense	16,770	0.1%	$ 20,000	****	20,000	1.2%
Energy (Utility) Expense	33,750	0.2%	$ 11,652	$ 0.41	35,680	2.1%
Administrative Expense (*****)	97,555	0.7%	$ 58,980	$ 0.72	101,175	5.8%
Repair & Maintenance Expense	14,820	0.1%	***	$ 0.30	17,581	1.0%
Rent Expense	180,000	1.2%	$ 180,000	$ –	180,000	10.4%
Depreciation – FF&E	24,000	0.2%	$ 24,000	$ –	24,000	1.4%
Depreciation – Leasehold improve.	24,000	0.2%	$ 24,000	$ –	24,000	1.4%
Depreciation – Smallware	24,000	0.2%	$ 24,000	$ –	24,000	1.4%
Total Operating Expenses	879,990	6.1%			1,006,514	58.0%

Operating Profits	13,079,905	90.6%	147,047	8.5%
Interest Expense (income)	18,000	0.1%	18,000	1.0%
Income before Income Taxes	13,061,905	90.4%	129,047	7.4%
Income Taxes (35%)	4,571,667		45,167	
Net Income (loss)	8,490,238	58.8%	83,881	4.8%

* Average check price has been adopted from the customer forecast (Table 6-01).

** Labor costs were adopted from the labor schedule total (Table 6-04).

*** The negative amount of the fixed cost is not used; and the variable cost per customer is adjusted.

**** Marketing expenses is adjusted to the same percentage (1.2%) of the revenues of the Y-2.

***** Administrative expenses include the Insurance of $18,000.

The pro forma Income Statement in Table 6.6 shows the results of the 2 years (Y-1 and Y-2) side by side to compare the operational results. The following is the explanation of how individual items have been forecast.

1. Revenues have been calculated by multiplying the average check price of $29.60 to the expected customer count of 58,604. According to our forecast, the customer traffic would increase by 15% in the Y-2 from the first year (Table 6.1). It was a naïve forecast, but in the beginning stage of a restaurant business, it is considered reasonable to achieve. The average check price was also naively estimated to increase by 10%. The average check price has been estimated to be $29.60. Since these calculations have been conducted using Excel, readers may come up with slightly different numbers. Personal discretions are recommended when there is small discrepancy. This increase is also acceptable to expect if R&B Grill tries to upsell their products with enhanced services. When these two numbers are multiplied, the expected revenues turn out to be $1,734,678 ($29.60 x 58,604). This result shows an acceptable increase of 20.1% from Y-1.

2. The Cost of Sales, as a true VC, was set at 33.5% of the revenues following the result of Y-1 to be conservative. In our forecast, the ideal cost was 32.41% (Refer to Table 5.2). The actual cost of sales percentage of 33.5% during the first year (Y-1) is quite close to the ideal condition. This result means that R&B Grill's inventory control system – purchasing and receiving, portion control, and waste control, and others – are implemented very efficiently. The company must maintain the same efficiency this way, if not better. When 33.5% of the food cost is applied to the expected revenues, the amount is $581,117 as presented in the pro format statement.

3. The Labor Expenses picked up the amount estimated through the employee work schedules as presented in Table 6.4. This expense is considered a mixed cost composed of a fixed amount and variable amounts. To accommodate unexpected situations, the labor cost includes $50,000 for contingency spending. The total labor cost will be 29.2% of the revenues. This is slightly higher than that of the Y-1 but still the overall percentage is relatively low. The prime costs are still maintained at 62.7% of the revenues in the forecast.

4. The Restaurant Operating Expense is also a mixed cost. It includes many different cost items. The breakdown of this cost into its FC and VC (refer

to Table 6.5) shows a negative FC (–$3,778). As pointed out, this is just the result of the calculation process of the regression analysis technique. Practitioners must use their own discretion to determine their most reasonable amount to forecast when statistical parameters turn out this way. In our forecasting, the negative FC amount is ignored. Instead, $1.25 has been applied to the per capita expense for each customer served. The result is $73,255 in the Y-2. Compared with the total revenues, it is 4.2%. This should be considered acceptable because it is almost the same as last year's percentage.

5. The Marketing Expense, in the breakdown process, shows a large amount of the FC and a negative amount of VC per customer (refer to Table 6.5). Due to the unique nature of this expense, the result like this should be explained. Simply, the result means that the marketing expense must be set at the highest amount. Out of the fixed highest amount, when the business serves more customers, the spending of this expense per customer will gradually go down. This is the interpretation of this outcome. In the budgeting process, it is recommended to set aside a fixed amount each month. In the second year (Y-2), it is expected that R&B Grill must do more to advertise their products and services. So, the marketing expenses are slightly increased to $20,000 for the year.

6. The next is Energy (Utility) Expense, which is also a mixed cost. The breakdown has yielded the FC amount of $971 per month, and an additional cost of $0.41 per customer served. These findings are directly applied to the total number of customers expected at the annual base. So, the entire amount of this expense is estimated at $35,680. When compared with the previous year's record, it is slightly lower than last year's 2.3%. This is acceptable considering that increased number of customers may and will enhance the energy efficiency.

7. Below the Energy Expense is the Administrative Expenses. This is also a composite expense item that includes many similar sub-expense accounts. As pointed out, this expense includes the Insurance Expense. This cost in Table 6.5 shows the FC amount of $4,915 per month and an additional cost that incurs per customer of $0.72. When these records are annualized and applied to the total number of customers, the result of the Y-2 is $101,175. The vertical analysis shows that it is 5.8% of the revenues. The same logic as in the Energy Expense applies to this situation. That is, increased number of customers enhances the cost efficiency.

8. The last mixed cost is the Repair and Maintenance. Table 6.5 shows the negative amount of the FC and $0.53 of additional cost per customer served. As we did in the case of the Restaurant Operating Expenses, the negative FC is ignored, and the additional cost per customer is adjusted to $.30, which has yielded $17,581 for the entire year. The amount is slightly higher than it was in Y-1, but the cost percentage remains the same. In the earlier days of the business, it is acceptable. However, as times go by, R&B Grill may have to increase its forecast in this expense due to the wear and tear of its properties.

9. The following items are all FCs – rent, depreciation, and interest expense. Annualized amount of the monthly expenses is applied.

10. The Income Tax is added in this forecast as the last expense to be safe and to be more conservative. Tax is neither a FC nor a VC. It is simply charged on the Income before Income Taxes (also known as Earnings before taxes, or EBT). The common tax rate of 35% is applied to the Income before Income Taxes.

11. Finally, after deducting all projected expenses, R&B Grill will earn $83,881 for the net profit. This will be added to the owners' equity. The projected net profit is 4.8% of the revenues. Its ROS is increasing according to this forecast.

As presented so far, revenues and expense are forecast based on the projected number of customers. It is reasonable because, in restaurant businesses, revenues and expenses are directly related to the number of customers served. In this context, the customers are the influencing factor (independent variable), and the revenues and expenses are the results (dependent variable). For each item (such as revenues and individual expense items), necessary parameters (intercept "a" and slope "b") have been identified using the Regression Analysis technique built in a spreadsheet program. In this process, profits (or losses) are calculated from the result of the forecast of the revenues and expenses. Profits are the result of the operations, and as such, they can only be estimated by forecasting the revenues and expenses first. The Table 6-07 below presents the progress of the business during the past two years in a summarized format. The table shows the changes of the company's assets, liabilities, and equity, supported by its operating activities. Major ratios are also presented.

Table 6-07 R&B Grill Ratio Analysis

Year	Y-0		Y-1		Y-2
Assets	$ 1,128,300		$ 1,135,705		$ 1,135,705
Liabilities	$ 475,500		$ 421,000		$ 421,000
Equity	$ 652,800		$ 714,705		$ 714,705
Revenues		$ 1,444,795		$ 1,736,026	
Oper. Profit		$ 80,505		$ 147,944	
E.B.T		$ 62,050		$ 129,944	
Net Profit		$ 21,877		$ 84,463	
Ratios					
Asset Turnover		1.28		1.53	
Return on Sales		1.5%		4.9%	
Return on Assets		1.9%		7.4%	
Return on Equity		3.2%		11.8%	

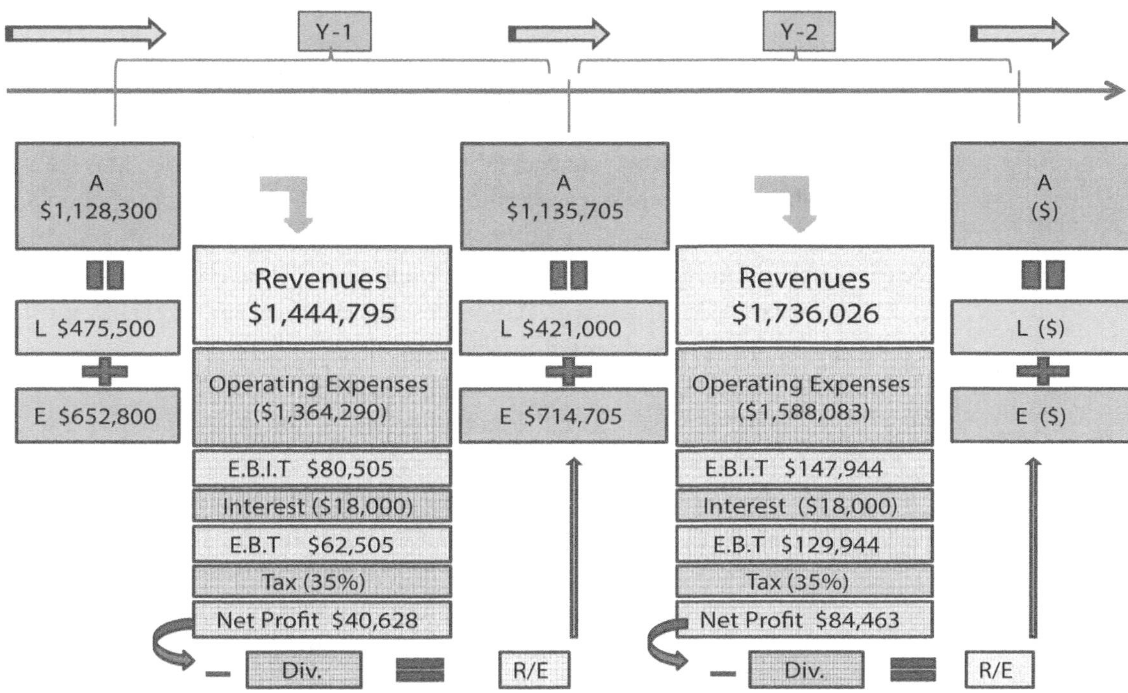

Figure 6.1 Business activities presented in time.

6.11 CONCLUSION

As mentioned earlier in this chapter, forecasting and budgeting activities provide initial goals. Once the goals are set, the management must try its best to achieve them. In their effort during the operations, the actual results must be compared frequently, and activities and plans must be adjusted to accommodate the realities. This is the internal controlling activities. The definition of internal control provided by the American Institute of Certified Public Accountants (AICPA) covers four objectives: safeguarding assets, checking accuracy and reliability of accounting data, promoting operational efficiency, and finally encouraging adherence to prescribed managerial policies. Mixing these four objectives into the issue of "controlling costs", it can be considered a series of activities of monitoring and (re)directing the course of operational actions to make sure that they are conducted to accomplish the established goals and objectives with the best efficiency.

Cost-control activities may mean checking the amount of profit on the bottom line. However, good practice of controlling costs must involve investigating every line of cost for discrepancies from the budgeted percentages of the revenues. If the food cost is different from the target percentage, the cause may be in portion control, or in purchasing price changes, or even changing patterns of customers' preference. Necessary corrections must be made to narrow down the future discrepancy. If labor cost shows unacceptable difference, the real cause may be with inefficient scheduling, or erroneous forecasting of customer counts, or even changing legal requirements on labor-related regulations. Again, appropriate adjustments must be made to the source of the discrepancy to narrow down the gap. All other lines of costs and expenses must be monitored and redirected this way as well.

For this reason, a well-established budget for every cost, as presented in this chapter, is required as prerequisite for good control systems. This is how to take care of pennies on costs to let the dollar of profits take care of itself. The more frequently this is conducted, the more efficiently the business will be handled toward the target goal. One common mistake is that many restaurateurs tend to become complacent without taking appropriate actions when their results turn out better than their goal. Relaxed control systems like this may put important business conditions in blind spots and may result in losing which otherwise could bring better results in the long run.

PART III

CASH FLOWS, PROJECTION, AND VALUATION

CHAPTER 7

THE CONCEPT OF CASH AND THE CASH FLOWS STATEMENT

7.1 INTRODUCTION

From the beginning of this book, it has been repeatedly pointed out that the ultimate goal of creating and running a business is adding value to the investments made by the owners. Now, it is necessary to discuss what is "adding value" and how it is measured. In accounting procedure, net profit generated from operations is returned to the owners. For this reason, creating profits is very important in any business and restaurants are no exception. However, net profit, due to its limitations, is not the true return to the owners. This chapter extends the discussion of the goal of a business beyond the concept of profit into the concept of cash flows.

7.2 PROFITS VERSUS CASH

First, profits are the result of the simple equation of revenues minus expenses (as in Revenues – Expenses = Profits). In a business, some expenses can be determined differently depending on the method adopted and they result in different amount of net profit. The Cost of Sales (food cost and beverage cost, in a restaurant business) is one example and depreciation expense is another. It is unnecessary to explain the technical details of individual methods in this book. To avoid confusion, however, all businesses must use the same method for calculating expenses consistently. For more details of specific methods of calculating these expenses, readers are recommended to refer to other books that are dedicatedly written on Financial Accounting. In this chapter, depreciation expense is emphasized to address its contribution to the concept of "value-adding." Depreciation, as a non-cash expense, does not involve the outlay of cash regardless of the amount determined. The business that reports depreciation keeps the amount of cash presented on the Income Statement.

Second, losses on the bottom line of the Income Statement do not mean that the business comes to an end. Oftentimes, businesses may have to go through hard times enduring poor operating performance with losses before they start growing successfully with profits. In extreme situations, businesses may continue its operations as long as they have enough cash despite losses from operations. Amazon.com is a good example. The company had reported losses continuously for long before it began generating profits. On the other hand, businesses that report significant amount of profits will collapse if they run out of cash. For these reasons, cash is practically more critical than profits. Hence, cash is often called the "life blood" of a business. This section introduces the difference between profit/loss and cash by itemizing the account titles explained in the earlier chapters. Then, the Statement of Cash Flows will be briefly introduced.

7.3 THE CONCEPT OF CASH FLOWS

A business may receive cash from different sources. The following examples point out how a business can obtain cash. When a business obtains cash, it is called Cash In-flows.

As in the example of R&B Grill, a business is created with the initial investments made by owners usually in cash. The business may take loans to raise necessary capital in cash. These are examples of financing activities of a business to obtain cash.

Once a business starts its operations, customers who are served usually pay their fees in cash. This is another source of cash the business obtains through operating activities. If a customer is allowed to pay their fees later, the unpaid amount is recorded in the Account Receivable for future collection. When these amounts are collected eventually, the business replenishes its cash. This is also considered as cash obtained from operating activities.

Finally, a business may raise the necessary cash occasionally by selling assets it owns such as a building, a line of business it operates, to name a couple. Selling (long-term) asset items to raise necessary funds is an investing activity. Usual investing activities of a business are purchasing necessary long-term assets such as building, equipment, and similar items. Thus, common investing activities cause cash to be spent (cash out-flows, in other words). It must be pointed out for readers not to mistake this investing activity for selling products to customers. Selling products is an operating activity that generates revenues. Selling assets is an "investing" activity of a business.

In summary, a business can obtain cash through three different types of activities – financing, operating, and investing – in the order of the explanation provided so far. Through the same activities, a business can also spend cash.

The usual investing activities, as introduced above, spend cash instead of obtaining it. The other two activities – financing and operating – also can spend cash. When a business pays cash dividends to the owners for their investments, it is a financing activity that spends cash. Another example of a financing activity that spends cash is spending for purchasing its own common stocks. When a public corporation buys back its own common stocks in the market, it is an example of a financing activity. As taking loans is a financing activity, paying for loans is also regarded as a financing activity. The only difference is paying for loans causes cash outflows. Cash outflows in operating activities will be explained later in this chapter.

As described, all three different types of business activities – operating, investing, and financing – cause cash either to flow in or to flow out. Since cash is the most important asset item considered the lifeblood of a business, the ability of a business to generate cash is the most important capability of the business to the investors and other stakeholders. In reality, the capability and the pattern of obtaining and spending cash is the key measurement criterion used by investors to assess the true value of a company. In this assessment, however, the Income Statement or the Balance Sheet does not deliver sufficient amount of information. The Income Statement only shows revenues and expenses, and the profit or loss as the result. The Balance Sheet, on the other hand, shows the cash balance and other information on the financial position. Neither report displays how or where cash has been obtained or spent. This limitation brings up the need for another report called the Statement of Cash Flows. The Statement of Cash Flows shows specific information regarding where and how cash has been obtained (simply, the source of cash) and where cash has been spent (the use of cash) by listing each type of activity –operating, investing and financing. The following section explains each segment with the examples of R&B Grill during Y-1.

7.4 CASH FLOWS FROM OPERATING ACTIVITIES

The Cash Flows Statement lists the operating activities on the top. Only through operating activities of serving customers, a business can grow itself to provide returns to the investors. To address the inflows and outflows of cash in operating activities, it is initially assumed that all revenues and expenses have been generated and incurred in cash. Accounts Receivable and Payable are ignored in the beginning but are separately assessed later and adjusted to pinpoint their impacts on cash.

With the initial assumption, the Net Profit (Loss) is considered the actual amount of cash obtained (spent) through operating activities. To this, noncash expenses (Depreciation and Amortization Expenses) are added because they remain in the business. Table 7.1 shows R&B Grill reported its net profit of $40,238 at the end of the Y-1. When the total depreciation expenses ($24,000 × 3 = $72,000) are added back to the net profit, the total cash inflows become $112,238. Although the net profit was only $40,238, the amount of cash generated from operating activities this year must be 112,238 when noncash expenses are added back. To this, the changes of current asset and current liabilities accounts are adjusted to explain their impacts on cash inflows and outflows through operating activities.

Table 7-01 The Income Statement (Y-1)

The Income Statement R&B Grill	Y-1
Revenues	1,444,195
Cost of Sales	484,300
Gross Profit	**959,895**
Operating Expenses	
Labor Expenses	401,800
Restaurant Operating Expenses	63,295
Marketing Expenses	16,770
Utility Expenses	33,750
Administrative Expenses	97,555
Repair & Maintenance Expenses	14,820
Rent Expenses	180,000
Depreciation – FF&E	24,000
Depreciation – Leasehold Improvement	24,000
Depreciation – Smallwares	24,000
Total Operating Expenses	879,990
Operating profits (EBIT)	**79,905**
Interest Expense (income)	18,000
Income before Income Taxes (EBT)	61,905
Income Taxes (35%)	21,667
Net Income (Loss)	**40,238**

7.4.1 Tax-Effect of Depreciation (as a Non-Cash, Tax-Deductible Expense)

It is important to notice the impact of Depreciation Expenses on the Income Tax. The Income Tax was estimated at 35% of the Income before income taxes (EBT; Earnings before tax). Without Depreciation Expenses, the taxable income would be $151,905 (current EBIT $79,905 + Total Depreciation $72,000). This EBIT without depreciation would have increased the taxable income (EBT) to $133,905. If this were the case, the Income Taxes would be $46,867 (← 35% of $133,905) compared with $21,667 as reported. By allocating $72,000 to the depreciation expenses, R&B Grill has reduced its taxable income to $61,905; and reduced the Income Taxes to $21,667. The depreciation expenses of $72,000 have saved R&B Grill $25,200 in the tax expense. This is one of the direct benefits of investing into a business. Besides, as mentioned, the entire amount of $72,000 remains in the company.

7.4.2 Working Capital Adjustment

The next is about the impact of the changes of Current Assets and Current Liabilities accounts on cash inflows and outflows. Inventory purchases and collecting Accounts Receivable are examples of this because they cause cash to be either spent or obtained. Additionally, changes of Accounts Payable and Accrued Expenses and other current liability accounts must also be explained for the same reason. By comparing their ending balances on the Balance Sheets from their beginning balances, their impacts must be identified. All these are considered as operating activities.

The details of these – Cash Flows from Operating Activities – are explained in the following section using Table 7.2. It shows the balances of individual accounts of the two consecutive accounting periods – the first month ("after 1 month") and the next year (Y-1). Readers are recommended to follow through each step and verify them in the Statement of Cash Flows presented in Table 7.3.

Table 7-02 The Balance Sheet (Y-1)

The Balance Sheet R&B Caterer, Inc.	After 1 month	Y-1
Assets:		
Current Assets		
Cash	438,000	640,225
Accounts receivable	60,000	40,000
Inventory	34,800	32,500
Prepaid Rent	85,000	-
Prepaid insurance	16,500	-
Total Current Assets	**634,300**	**712,725**

Table 7-02 (Continued)

Long-term Assets		
FF&E	150,000	150,000
less Acc. Depreciation	(2,000)	(26,000)
Leasehold Improvement	250,000	250,000
less Acc. Depreciation	(2,000)	(26,000)
Smallware	100,000	100,000
less Acc. Depreciation	(2,000)	(26,000)
Net FF&E	494,000	422,000
Total Long-term Assets	494,000	422,000
Total Assets	**1,128,300**	**1,134,725**
Liabilities & Equity		
Liabilities		
Current Liabilities		
Accounts payable	55,000	109,000
Accrued expenses	21,480	93,667
Unearned revenues	-	-
Total Current Liabilities	76,480	202,667
Long-term Liabilities		
Notes payable-A	150,000	130,000
Notes payable-B	250,000	110,000
Total Long-term Liabilities	400,000	240,000
Total Liabilities	**476,480**	**442,667**
Owners' Equity		
Common stock		
- Rachel & Brad	450,000	450,000
Common stock		
- Jane	200,000	200,000
Retained Earnings	1,820	42,058
Total Owners' Equity	**651,820**	**692,058**
Total Liabilities & Equity	1,128,300	1,134,725

7.4.2.1 Net Income of Y-1

As mentioned, the information of cash flows starts with the operating results under the assumption that all revenues and expenses have been collected and disbursed in cash. Under this assumption, the Net Income is considered the initial amount of cash obtained from operations.

7.4.2.2 Plus: Non-Cash Expenses

To the Net Income, the total amount of non-cash expenses must be added because these are not paid in cash to anyone and the amount of cash remains in the company. In practice, non-cash expenses include many other items such as amortization and others of a similar nature. Most businesses provide a long list of items in this section.

The next part is the "Working Capital Adjustments," which explains the increase and decrease of cash through the business activities that support operations. By comparing the balance of each account that belongs to the Current Assets and the Current Liabilities, the impact on cash is determined individually. As the current assets and the current liabilities are often called Working Capital, this section is so titled.

7.4.2.3 Accounts Receivable

This account shows the amount of cash not collected at the time of the transaction of the revenues. If the balance of this account increases, the impact is a decrease in cash because the company has failed to replenish its cash. On the other hand, if it decreased, the impact is an increase of cash because outstanding amounts have been collected. R&B Grill's Accounts Receivable in its Balance Sheet records shows a decrease of $20,000 from $60,000 in the beginning to the $40,000 at the end. R&B Grill has collected cash from its outstanding Accounts Receivable. Thus, cash has increased. For this reason, the decrease of Accounts Receivable represents the increase of $20,000 in cash flows.

7.4.2.4 Inventory

Inventories are constantly bought and used in restaurant operations. Whenever they are purchased, cash outflows due to the payment. When the ending balance of inventory is larger than the beginning balance, it is considered as more purchase, thus cash outflows for the payment. If it decreases, however, it is considered as less purchase of inventory, which must have saved cash. When cash is saved, its balance increases. The ending inventory of R&B Grill is $32,500, which is smaller than the beginning balance by $2,300. R&B Grill has saved cash by purchasing inventory less by $2,300 thus cash increased (or saved). This impact is presented in the Statement of Cash Flows (Table 7.3).

7.4.2.5 Prepaid Rent and Prepaid Insurance

These are prepaid expenses. They are deposits made for specific purposes. R&B Grill uses these deposited funds for relevant expenses – rent or insurance, in this case – when they become due. If the balance of Prepaid Expenses grows over time on the balance sheet, more deposits must have been made in cash during the accounting period. Thus, cash has decreased. When they decrease, it means that the business has used them for specific expenses rather than disbursing its cash for payments. Thus, cash on hand is saved (increased) at the time of payment transaction.

As the balance of both of these two accounts – Prepaid Rent and Prepaid Insurance – of R&B Grill during Y-1 decreased, the result indicates that cash has been saved during the year. Differently put, R&B Grill did not use its cash on hand for these expenses. Instead, it used its deposits that had been made in advance. Hence, these two accounts show positive amount indicating cash flows has increased in Table 7.3.

R&B Grill's current asset accounts are adjusted for their impacts on the company's cash flows. It can be summarized that when a current asset account increases its balance in one accounting period, its impact on cash is negative. The opposite is true with the decrease of a current asset account balance. The decrease of asset account balance impacts cash flows positively. They are just like purchasing or selling of asset items. When an asset item is purchased, the balance of the asset account increases; but cash must decrease due to the payment. When an asset item is sold, its balance decreases but cash must be obtained. The next part is about the impact of current liabilities on cash. Readers are recommended to verify the information on the Statement of Cash Flows (Table 7.3) while reading the following explanation.

7.4.2.6 Accounts Payable

This account records the unpaid amount of purchases. Because the payment has not been made for purchases, the amount of cash – that has not been paid – is still retained in the company. So, it results in an increase of the cash balance. On the other hand, the decreasing balance of this account means that the company has made payments in cash. Thus, cash balance decreases. The Balance Sheet of R&B Grill shows the Accounts Receivable has increased its balance by $54,000 from $55,000 to $109,000. This means that the business kept cash of $54,000 that should have been paid during the year, which has a positive impact on cash balance.

7.4.2.7 Accrued Expenses

Accrued Expense accounts are also liabilities that represent unpaid amounts of expenses that have been recorded in the Income Statement. They work exactly the same as the Accounts Payable account as explained above. R&B Grill shows the increase of this account balance by $72,187 from $21,480 to $93,667 at the end of Y-1. This indicates that the business has been keeping cash instead of paying $72,187 for relevant expenses. As the result, its cash balance must have increased.

7.4.2.8 Unearned Revenues

This is another current liability account. An increased amount of this account means that the company has received cash prior to providing services. Because the company has taken cash in advance of services rendered, its cash flows have increased. The decrease of this account balance means the company has rendered its services for the cash received. It is considered just like paying for the debt. As a liability account, it works just like other liability accounts.

In summary, the current liabilities work exactly the opposite way to that of the current assets. The increase of current liabilities will have positive impacts on cash while the decrease will negatively impact the cash balance. Table 7.3 presents the Cash Flows Statement of R&B Grill for Y-1. Now that the cash flows from operating activities are completed with the explanation of the net profit, depreciations, followed by the adjustments of the working capital accounts, the next step is to explore the impact of cash flows caused by the investing and financing activities of the business.

Table 7-03 The Cash Flows Statement (Y-1)

The Statement of Cash Flows The R&B Caterer, Inc. For the year of Y-1	
Cash Flows from Operating Activities	**Inflows (Outflows)**
Net Profit of the Y-1	40,238
Plus: non cash expenses	72,000
Working Capital Adjustments	
Accounts receivable	20,000
Inventory	2,300
Prepaid Rent	85,000
Prepaid Insurance	16,500
Accounts payable	54,000
Accrued expenses	72,187
Unearned revenues	-
Cash flows from investing activities	
FF&E	-
Leasehold improvement	-
Smallware	-
Total cash flows from investing activities	**-**
Cash flows from financing activities	
Notes payable – A	(20,000)
Notes payable – B	(140,000)
Common Stock – Rachel & Brad	-
Common Stock – Jane	-
Total cash flows from financing activities	**(160,000)**
Net Cash Flows	**202,225**
Beginning Cash	438,000
Net cash flows	202,225
Ending Cash	640,225

7.5 CASH FLOWS FROM INVESTING ACTIVITIES

Investing activities refer to purchasing or selling long-term assets. When a company invests into its long-term assets, the balance of long-term assets grows in the Balance Sheet. This is considered as cash outflows that decrease cash balance. During this process of estimating cash flows, liabilities involved in the investing activities are ignored because liabilities increases will be separately analyzed at the section of financing activities. The mechanism of this analysis remains the same as those introduced in the current asset items in the operating activities. The accumulated depreciation of individual long-term asset items is not included in this process, either. Instead, only the purchase cost is used.

R&B Grill shows, in the Long-term Assets section of the Balance Sheet, three different investments made into its long-term asset items – FF&E, Leasehold Improvements, and Smallwares. There are no changes in the original values of these items, which indicate that, during the Y-1, R&B Grill has made no additional investments. It has not disposed any of existing long-term assets, either. With no new investments or no disposals, there are no impacts on its cash flows from investing activities. In other words, cash has been neither spent nor obtained. Hence, this section shows zeros (0's) in the Cash Flows Statement (Table 7.3). If there have been long-term investments, their impacts on the cash flows will be negative indicating outflows of cash and vice versa.

7.6 CASH FLOWS FROM FINANCING ACTIVITIES

Financing activities refer to the method of obtaining necessary funds – either through issuing long-term debt or through issuing equity. When a business issues new long-term debt, it brings in cash. When it issues new common stocks to raise necessary capital, it also brings in cash. Thus, both activities will result in increasing its cash inflows. On the other side, financing activities also involve the payment of long-term debt, the payment of dividends, or the purchase of the company's own stocks. All these activities will decrease the cash balance. The section of the financing activities of the Cash Flows Statements introduces the impacts of these activities.

R&B Grill's Balance Sheet shows the decrease of its long-term debts (both in Notes Payable A & B). Notes Payable-A decreased by $20,000;

while Notes Payable-B also decreased by $140,000. These decreases are the result of the payments of the debt throughout the year. These payments are considered the use of cash. As they decrease cash balance, their impacts are shown as negative numbers in this section.

7.7 CONCLUSION OF THE CASH FLOWS STATEMENT

All these three-dimensional activities – operating, investing, and financing – either bring in or disburse cash. The bottom line of the statement titled "Net Cash Flows" is the total of the cash flows from all three activities. R&B Grill has increased its cash amount throughout the entire year of Y-1 by $202,225. More specific summary is as follows:

1. It generated $362,225 in cash through its operating activities.
2. It did not make any new investments during the year. It did not sell or get rid of any long-term assets.
3. Finally, it spent $160,000 in cash to pay down its long-term debt. There has been no common stock issued during the year. Nor were dividends paid out.
4. As the result, the company has obtained additional cash for $202,225 during the year of Y-1, which is presented as the Net Cash Flows on the bottom.

The last three lines show the change of R&B Grill's cash from the beginning of Y-1 to the end of the year. It started with $438,000 in its cash account. During the entire year, the business has increased its cash by $202,225 and the ending balance is $640,225. As pointed out earlier, it is the Statement of Cash Flows that presents the activities of cash inflows and outflows for their specific sources and uses.

7.8 ADDITIONAL INFORMATION ON DETERMINING THE VALUE OF A BUSINESS USING ITS CASH FLOWS INFORMATION

One frequently mentioned term related to cash flows is the Free Cash Flows. This information is not presented in the Cash Flows Statement. It refers to the difference between the Cash Flows from operating activities and those from the investing activities. A business must make continuous investments to sustain its current operations by spending necessary amounts of capital expenditure. These are considered investing activities that use cash. As the business must use the cash generated from its operating activities, this capital expenditure must be taken out of the cash obtained from operating activities. The remaining amount is called the Free Cash Flows because it can be used by the management for any purpose. For this reason, the Free Cash Flows are often used in the valuation process of a business. This information is used to estimate the market value of a business.

The fundamental accounting equation of $A = L + E$ applies not only to the book value but also the market value. The only market value presented in the Balance Sheet is the amount of Total Liabilities. All others are in book value. Once the market value of the entire assets is estimated, the debt amount (Total Liabilities) is taken out to estimate the market value of the Owners' Equity. This is a brief introduction of how valuation process is conducted. More details of this will be explained in Chapter 10 on Valuation. In case of R&B Grill for the year of Y-1, it did not spend cash for in investing activities. Therefore, its Free Cash Flows is the same as its Cash Flows from Operating Activities of $362,225. Out of this amount, the company has paid its long-term debt for $160,000. The Net Cash Flows presented on the bottom of the report is the leftover amount after taking the payment of the long-term debt of the year. Technically, the "Net Cash Flows" is the true cash amount that can be returned to its owners.

CHAPTER 8

LONG-TERM PROJECTION OF A BUSINESS

8.1 INTRODUCTION

This chapter introduces a long-term forecasting model that is a stepping-stone to estimate the market value of a business. In Chapter 6, using a naïve method of forecasting, the annual customer count of the Y-2 was projected to increase 15% to 58,604 and the average spending of each customer (average check price (ACP)) was projected to increase 10% to $29.60. If these projections work, the annual sales of Y-2 will be $1,734,737. These are presented again in Table 6.1. All expenses were estimated based on the projected customer count using their fixed cost amounts and the variable cost amount per customer as the relevant statistical parameters obtained through the Regression Analysis technique. Finally, ratio analysis was conducted, and the results were examined with those of well-known similar businesses to test the reasonableness. It was concluded that the forecast of Y-2 was reasonable and achievable (refer Table 6.1: Customer Count and Sales Projection).

This chapter introduces long-term forecasting by adjusting necessary operating elements that determine future revenues and major expenses. This process must adopt reasonable scenarios regarding possible changes in the influential factors, such as the customers, competitors, supply chain, and regulations, which are believed to affect the business performance directly. For example, the 15% growth of the customer count used for Y-2 may be achieved in the second year only because the customer count of the first year was small. It will not be reasonable to expect to achieve the same growth rate in the following years. This is the limitation of naïve methods. The maximum number of customers must be determined by addressing the physical size and the capacity of the business. R&B Grill has 100 seats. It is also unthinkable to fill up every seat of each table during the operating hours. If each table accommodates two customers, only 50 seats will be used to generate revenues. Here comes the concept of the ideal number of seat turnover.

8.2 CONSERVATIVE PROJECTION OF REVENUES BASED ON THE PHYSICAL SIZE OF R&B GRILL

According to the projection of the Y-2 (as introduced in Table 6.1), the estimated average weekly customer count of R&B Grill was 1,127. The weekly seat turnover, according to this estimate, is slightly higher than 11 (1,127/100). The average daily seat turnover is lower than 2 (1.61 to be exact). Taking into account the peak on weekends, this would be considered reasonably achievable forecast.

The limitation of the naïve method can be supplemented by developing multiple scenarios. The following two tables (Tables 8.1(A) and 8.1(B)) present revised customer count projections by the hour of each operating day in two different scenarios. The first one (8.1(A)) is for the Y-2 according to the minimum scenario to be more conservative and the next (8.2(B)) is the maximum scenario for the year when the business fully grows into its mature stage. It is expected that the revenues of R&B Grill will grow from the estimated result of Y-2 (as presented in the Table 8.1(A)) into those presented in Table 8.1(B). It must be noticed that the average weekly customer count in the minimum scenario is slightly increased to 1,133. The rationale of these scenarios is introduced in the following section using necessary operational analysis.

According to the estimate in Table 8.1(A), during the second year (Y-2), the seat turnover ranges from the lowest of 0.86 on Sundays to the highest of 3.23 on Saturdays. During the busiest hour on Saturdays from 7 p.m. to 9 p.m., R&B Grill will accommodate fewer than 70 customers in its 100-seat dining hall, which should be considered reasonable. By filling up its tables by 70%, R&B Grill reaches its near maximum capacity on Saturdays. The only way to increase its revenues by attracting

Table 8-01(A) Customer Count Projection – Minimum Scenario

Hours	Mon	Tue	Wed	Thu	Fri	Sat	Sun	Total
3:00–4:00							6	6
4:00–5:00	15	10	12	16	20	24	10	107
5:00–6:00	16	14	18	20	30	34	20	152
6:00–7:00	24	26	26	32	46	56	20	230
7:00–8:00	26	25	24	30	64	64	14	247
8:00–9:00	24	22	22	20	40	64	10	202
9:00–10:00	10	8	16	18	30	56	6	144
10:00–11:00					20	25		45
Total Customers	115	105	118	136	250	323	86	1133
Available seats	100	100	100	100	100	100	100	
Seat Turnover	1.15	1.05	1.18	1.36	2.5	3.23	0.86	11.33

Customer count projection for the minimum scenario	
Annual customer count (52 weeks): 15% increase from the Y-1	58,916
Average Check Price (Adjusted per expected value: Table 8-02)	$ 26.35
Annual revenues (Adjusted with the revised Average Check Price)	$ 1,552,437

Table 8-01(B) Customer Count Projection – Maximum Scenario

Hours	Mon	Tue	Wed	Thu	Fri	Sat	Sun	Total
3:00–4:00							5	5
4:00–5:00	15	10	10	18	20	20	8	101
5:00–6:00	18	15	25	20	40	40	20	178
6:00–7:00	35	28	35	50	56	56	18	278
7:00–8:00	32	30	40	55	64	64	14	299
8:00–9:00	25	23	25	40	58	58	10	239
9:00–10:00	10	8	15	15	54	54	4	160
10:00–11:00					20	20		40
Total Customers	135	114	150	198	312	312	79	1300
Available seats	100	100	100	100	100	100	100	
Seat Turnover	1.35	1.14	1.5	1.98	3.12	3.12	0.79	13

Customer count projection for the maximum scenario	
Annual customer count (52 weeks x 1,300)	67,600
Average Check Price (Best Case Scenario)	$ 29.00
Annual revenues	$ 1,960,400

more customers is to utilize other days. Fridays as the second busiest day, according to the projection, still has some room to increase its seat turnover. Other weekdays also have enough rooms, but it will be difficult for a restaurant to bring in a large number of customers during weekdays. Many restaurateurs offer a variety of incentive plans such as happy hour meals or discounts on specific items to increase customer counts. However, cost-benefit analysis must be conducted thoroughly before providing this type of incentives for their long-term sustainability. Labor scheduling must be reviewed for such projection to prevent possible bottleneck problems in services. The payroll projection based on the labor scheduling will be introduced later.

Table 8.1(A) presents the minimum customer count based on the result of Y-1. The next question is "what is the maximum number of customers when R&B Grill fully grows into the full capacity?" Table 8.1(B) presents the estimated customer count of the maximum stage. In our scenario, it is assumed that R&B Grill will reach its mature stage in the seventh year (Y-7). It is assumed to accommodate 1,300 guests per week, which will be average daily seat turnover of 1.85. According to this estimate, the daily seat turnover ranges from the lowest of 0.79 on Sundays to the highest of 3.12 on Fridays and Saturdays. Looking into the hourly breakdown, neither days of the weekend will experience the overflows of the guests. One thing to remember is that some customers who enter in one-hour slot may stay over to the next hour. This factor is presented in the "Labor Scheduling and Cost" table (Table 8.4(B) Wage Projection – max), which shows the restaurant still has a couple of table empty for extra customers.

8.3 PER CAPITA SPENDING (AVERAGE CHECK PRICE: ACP) ESTIMATE

Once the customer count is estimated, the next is the Per Capita Spending, or the ACP. This was also forecast to increase by 10% to $29.60 using a naïve method. This estimate needs more analysis and adjustments. According to the 2015 annual reports of national chain restaurants, the amount of the ACP of such brands as the Outback Steakhouse Restaurants (Bloomin' Brands) and similar companies was around $23 to $26. If R&B Grill wants to pursue the goal as forecast at $29.60, which is higher than the nationally well-known chains by $3 to $6, Rachel and Brad must have very good reasons to believe it. The reasons that support their confidence may include, but are not limited to, the economic/financial conditions of its clientele in the region, superior quality of their products and services, strong supply chains that will stay constant, unchallengeable loyalty of its customers, to name only a few. The competition in the area is also a very important factor to take into consideration. There are many other possibilities that may negatively affect the result. To be more flexible to accommodate uncertainties, another approach can be made by constructing and mixing different scenarios.

8.4 PROBABILISTIC METHOD TO ESTIMATE THE MOST REASONABLE ACP

The 10% increase in the ACP was a naïve conclusion. It may be possible; or it may even be overachieved with remote possibility. Most likely, however, it may not be achieved. The ACP of the first year (Y-1) turned out to be around $27 ($26.93, to be exact). If the business continuously satisfies the customers, it may be able to maintain the first year's ACP, or even increase it. However, it is not recommendable to expect that the customers will continuously spend higher than the national average amount for R&B Grill's products and services. Rachel and Brad can prepare a simple set of scenarios as follows:

1. Scenario 1 (Optimistic): The ACP will increase as expected to $29.62. The probability of this happening is about 15%. The probability is the manager's intuitive decision.

2. Scenario 2 (Moderate): The ACT will remain at the same as in the Y-1, at $26.93 with no increase. The probability of this happening is about 60%. The probability is determined using the example of national chain restaurant companies' ACT.

3. Scenario 3 (Pessimistic): The ACT will decrease to $23, which is the lowest level at nationally well-known chains. The probability of this happening is 25%.

The three different scenarios need to be combined into one consolidated result to project the ACP of the second year (Y-2), which is presented in Table 8.2. According to the projection with the relevant probabilities, the second year's AVP is forecast to be $26.35, which is 2.15% lower than that of the first year. This newly obtained information will be used to revise the Pro forma Income Statement of Y-2.

Table 8-02 Expected Value of the Average Check Price (ACP)

Scenario	Target ACP	Probability	Expected ACP
1 (Optimistic)	$ 29.62	15%	$ 4.44
2 (Moderate)	$ 26.93	60%	$ 16.16
3 (Pessimistic)	$ 23.00	25%	$ 5.75
Weighted Average Check Price		100%	$ 26.35
a: This is 2.15% reduced price from the Y-1.			
b: Average weekly customer Count (per Table 8-01a)			1,133
c: Annual customer count (52 weeks)			58,916
c: Annual revenues (ACP x Annual Customer Count)			$ 1,552,437

Combining the projections of the customer count and the ACP, the annual revenues of Y-2 will be $1,552,437. This is the result of serving 1,133 customers per week, with the ACP at $26.35 for 52 weeks. In the previous annual budget of Y-2, revenues were projected to be $1,736,026. The revised forecast presents the annual revenues lowered by almost 10.5% or $183,589. This reduction is the result of adjusting the ACP accommodating the potential competitors' influence. This should be considered more conservative projection. Table 8.3 presents the revised operational results of Y-2.

Table 8-03 Pro Forma Income State of the Y-2
(Revised with the lower Average Check Price)

The Pro Forma Income Statement of Y-2 R&B Grill Y-2 Forecast (per Tables 8-01a & 8-02)	FC		VC/cust		Y-2	V.A. %
	Customer #			58,916		
Revenues	Avg. Check*		$	26.35	1,552,437	100.0%
Cost of sales	Cost %			33.5%	520,066	33.5%
Gross Profit					1,032,370	66.5%
Operating Expenses	FC (annual)		VC/cust			
Labor expenses	per labor schedule**				507,395	32.7%
Restaurant operating expense	***		$	1.25	73,645	4.7%
Marketing expenses	$	20,000		****	20,000	1.3%
Energy (Utility) expenses	$	11,652	$	0.41	35,808	2.3%

Table 8-03 (Continued)

Administrative expenses (*****)	$	58,980	$	0.72	101,400	6.5%
Repair & maintenance expenses		***	$	0.30	17,675	1.1%
Rent expenses	$	180,000	$	–	180,000	11.6%
Depreciation – FF&E	$	24,000	$	–	24,000	1.5%
Depreciation – Leasehold improve.	$	24,000	$	–	24,000	1.5%
Depreciation – Smallware	$	24,000	$	–	24,000	1.5%
Total Operating expenses					1,007,922	64.9%
Operating profits					24,448	1.6%
Interest expense (income)					18,000	1.2%
Income before income taxes					6,448	0.4%
Income taxes (35%)					2,257	
Net Income (loss)					4,191	0.3%

* Average check price has been revised using the Expected Value method (Table 8-02).

** Labor costs were adopted from the labor schedule total (Table 6-04).

Balance Sheet records – projection – the R&B Grill's financial positions will be maintained at the current values.

Average Assets	1,135,705
Average Liabilities	421,000
Average Equity	714,705

Ratio Summary – projection	Y-2
Asset Turnover	1.37
Return on Sales	0.27%
Return on Assets	0.37%
Return on Equity	0.59%

The revised pro forma Income Statement of Y-2 in Table 8.3 shows the result with almost no profits (0.3%) on the bottom line due to the ACP reduction. All expenses remain the same as shown before because customer count has not changed. The best way R&B Grill can enhance its operating result is either to increase its ACP or to increase its customer traffic. It is the operators' task to design their own plans to achieve this goal.

The result of R&B Grill's projected performance of Y-2 is also presented in a few key financial ratios on the bottom of Table 8.3. With the reduced ACP, the asset turnover is 1.37 with the very low results in other profitability ratios. Almost at the break-even point, this could be considered the bare minimum results. As mentioned earlier, if the business increases the ACP higher than the currently estimated level or increase the customer counts, the operational result will gradually grow better. For example, the Asset Turnover ratios of major chain restaurants are around 1.8 to 2.0. Comparing with these trends, R&B Grill, with its Asset Turnover at 1.37, still has a potential to grow its revenues. When the ACP was naively forecasted to be $29.62, the revenues were projected to be over $1.7 million. This could have achieved its Asset Turnover at 1.5. One dilemma R&B Grill faces is the physical size of its property that has only 100 seats. Most successful casual-theme full-service national chain competitors have at least 180 to over 200 seats in their properties. With a larger property size, it becomes easier to serve larger number of customers, which will result in better asset turnover. This is the main concept of the Economy of Scale. To overcome this physical limitation, R&B Grill must develop its strengths in the quality of its products and services, so that customers will spend premium amounts. This strategy will require better-quality ingredients with frequent menu development, continuous investments into its physical facility to maintain or enhance the overall ambiance, and ceaseless training for its employees to satisfy customers' expectation.

8.5 LONG-TERM PROJECTION

The next part introduces the long-term projection based on the adjusted assumption that R&B Grill implements all necessary plans and they work reasonably well as planned from Y-2. The following is the summary of the forecast of that year:

- First, the revenues were readjusted to $1,552,437 based on the reduced ACP of $26.35.
- The cost of sales remains at 33.5% of the revenues, which is the ideal cost percentage. It needs be controlled at this level. Any significant variation from this percentage must be thoroughly analyzed for its causes and be readjusted frequently.

- Labor cost is readjusted using the labor-scheduling table (refer to Tables 8.4(A)–8.4(C)).

- All other costs are estimated based on the customer count using the result of the regression analysis introduced in the Chapter 6.

With the Pro forma Income Statement of the Y-2 prepared as explained so far, it is necessary to project the long-term operations. The long-term projection can be considered the road map to reach the maximum stage of the operations where the revenues are generated with the full utilization of the resources. As mentioned earlier, a restaurant business has its physical limitation in serving its guests. This limitation must be projected reasonably and all necessary costs and expenses also must be assessed. Only when the expected revenues surpass the necessary amount of the expenses, the business can be justified for the investments that enable the operations to continue. The long-term projection of R&B Grill is based on the following assumptions and mechanism:

1. Revenues will grow from the second year (Y-2) as presented in the pro forma Income Statement to the seventh year (Y-7) when it reaches its maximum stage and mature. This is achieved by around 2% growth each year. Annual grow rate of 2% is estimated based on the average inflation rate of the past decade.

2. The Cost of Sales is set at 33.5% of the revenues, although the ideal cost was determined at 32.4% in Tables 5.2. It was increased to 33.5% to be flexible to accommodate necessary fluctuations.

3. Labor cost is prepared based on the labor schedules and payroll estimate presented in Tables 8.5(A)–8.4(C). Labor schedules are prepared to accommodate customer traffic. In case of unexpected labor cost needs, a separately budgeted Contingency Labor Cost (introduced in the Table 8.4(C)) will be used to absorb the additional expenses without hurting the operations' bottom line. According to the schedule, labor cost will gradually reduce from 32.9% in Y-2 to 28.6% in Y-7 as revenues continuously increase to the expected level. The growth rate of labor cost is lower than the 2% per year, which will have positive impacts on the bottom line. Tables 8.4(A) and 8.4(B) introduce the labor schedule of the front-of-the house for the slowest year (Y-2) and the busiest year (Y-7). Finally, Table 8.4(C) shows the consolidated annual labor costs

Table 8-04(A1) Daily Wage Projection – Front of the House – Y2 (minimum)

Monday Projection					
Hours	Cust.*	In hall**	Server hours***		Total Hours
3:00–4:00			2		2
4:00–5:00	15	23	2		2
5:00–6:00	16	28	2	2	4
6:00–7:00	24	37	2	2	4
7:00–8:00	26	38	2	2	4
8:00–9:00	24	29	2	2	4
9:00–10:00	10	10	1	2	3
10:00–11:00				2	2
Total	115				
# of hours			13	12	25
Wage per hour – tipped employee					$ 5.50
Total daily wage					$ 137.50

Friday Projection					
Hours	Cust.*	In hall**	Server hours***		Total Hours
3:00–4:00			3		3
4:00–5:00	20	35	3		3
5:00–6:00	30	53	3	3	6
6:00–7:00	46	78	3	3	6
7:00–8:00	64	84	3	3	6
8:00–9:00	40	55	3	3	6
9:00–10:00	30	40	2	2	4
10:00–11:00	20	20	2	2	4
11:00–12:00				2	2
Total	250				
# of hours			22	18	40
Wage per hour					$ 5.50
Total daily wage					$ 220.00

Tuesday Projection						
Hours	Cust.*	In hall**	Server hours***		Total Hours	
3:00–4:00			2		2	
4:00–5:00	10	17	2		2	
5:00–6:00	14	27	2	2	4	
6:00–7:00	26	38	2	3	5	
7:00–8:00	25	36	2	3	5	
8:00–9:00	22	26	2	2	4	
9:00–10:00	8	8		2	2	
10:00–11:00				2	2	
Total	105					
# of hours			12	14	26	
Wage per hour					$	5.50
Total daily wage					$	143.00

Saturday Projection						
Hours	Cust.*	In hall**	Server hours***		Total Hours	
3:00–4:00			4		4	
4:00–5:00	24	41	4		4	
5:00–6:00	34	62	4	3	7	
6:00–7:00	56	88	4	3	7	
7:00–8:00	64	96	4	3	7	
8:00–9:00	64	92	4	3	7	
9:00–10:00	56	68	4	3	7	
10:00–11:00	25	25		3	3	
11:00–12:00				3	3	
Total	323					
# of hours			28	21	49	
Wage per hour					$	5.50
Total daily wage					$	269.50

Wednesday Projection

Sunday Projection

Table 8-04(A1) (Continued)

Hours	Cust.*	In hall**	Server hours***		Total Hours
3:00–4:00			2		2
4:00–5:00	12	21	2		2
5:00–6:00	18	31	2	3	5
6:00–7:00	26	38	2	3	5
7:00–8:00	24	35	2	3	5
8:00–9:00	22	30	2	3	5
9:00–10:00	16	16		3	3
10:00–11:00				2	2
Total	118				
# of hours			12	17	29
Wage per hour					$ 5.50
Total daily wage					$ 159.50

Hours	Cust.*	In hall**	Server hours***		Total Hours
2:00–3:00			1		1
3:00–4:00	6	11	1		1
4:00–5:00	10	20	1	2	3
5:00–6:00	20	30	1	2	3
6:00–7:00	20	27	1	2	3
7:00–8:00	14	19	1	2	3
8:00–9:00	10	13		2	2
9:00–10:00	6	6		2	2
10:00–11:00				2	2
Total	86				
# of hours			6	14	20
Wage per hour					$ 5.50
Total daily wage					$ 110.00

Thursday Projection

Hours	Cust.*	In hall**	Server hours***		Total Hours	
3:00–4:00			2		2	
4:00–5:00	16	26	2		2	
5:00–6:00	20	36	2	2	4	
6:00–7:00	32	47	2	2	4	
7:00–8:00	30	40	2	2	4	
8:00–9:00	20	29	2	2	4	
9:00–10:00	18	18		2	2	
10:00–11:00				2	2	
Total	136					
# of hours			12	12	24	
Wage per hour					$	5.50
Total daily wage					$	132.00

* The customer counts are the ones that leave by the end of each hour.

** In-hall customer count of each hour is estimated with a half of the next hour customer added to the current ones. For example, during the hour of 6-7, 17 customers are finishing their meal, and 11 additional customers are being served.

*** Server hours stand for the number of servers working in each hour. For example, from 5 to 6 o'clock, 4 servers are scheduled to work to serve 28 customers in the dining hall. During this one hour, 4 hours' payroll is being incurred.

More details are introduced in the text.

Table 8-04(A2) Weekly Wage Total – Front – Y2 (minimum)

Day	Hour	Wage	Amount
Monday	25	$ 5.50	$ 138
Tuesday	26	$ 5.50	$ 143
Wednesday	29	$ 5.50	$ 160
Thursday	24	$ 5.50	$ 132
Friday	40	$ 5.50	$ 220
Saturday	49	$ 5.50	$ 270
Sunday	20	$ 5.50	$ 110
Weekly Total	213		$ 1,172
Annual Total (52 wks)			$ 60,918

* These are tipped employees.

of the minimum scenario of Y-2 and the maximum scenario of Y-7. The consolidated labor cost (Table 8.4(C)) also includes the amount of salaries, reserved amount for contingency labor expenses, and labor-related expenses. Labor-related expenses include Payroll Taxes, Insurance, benefits, and other employee-related expenses.

4. All other expenses are supposed to increase according to the customer count growth. These increases are projected using the fixed cost and variable cost amounts of each expense obtained using the regression analysis of the first year (Y-1). As explained earlier, negative fixed costs are replaced with zero (0); and only the variable cost amount per customer is used to remain conservative in the projection of those expenses that showed negative fixed costs in Y-1. Marketing expenses are treated as a fixed cost. It is reasonable that the percentage of the marketing expenses shrinks as the revenues grow in time.

The result of the projection (as introduced above) is presented in the following Table 8.5 "Long-Term Projection." It is assumed that R&B Grill reaches its maximum capacity in the seventh year (Y-7), and it is expected to continue its mature stage into the further future.

Below the projected Income Statement of the next 6 years (Y-2 through Y-7) is the summary of major financial ratios. It is also assumed that R&B Grill will maintain its financial position of the beginning of the

Table 8-04(B1) Daily Wage Projection – Front of the House – Y7 (maximum)

Monday Projection					
Hours	Cust.*	In hall**	Server hours***		Total Hours
3:00–4:00			2		2
4:00–5:00	15	24	2		2
5:00–6:00	18	35	2	2	4
6:00–7:00	35	51	2	2	4
7:00–8:00	32	44	2	2	4
8:00–9:00	25	30	2	2	4
9:00–10:00	10	10	1	2	3
10:00–11:00				2	2
Total	135				
# of hours			13	12	25
wage per hour – tipped employee					$ 5.50
Total daily wage					$ 137.50

Friday Projection					
Hours	Cust.*	In hall**	Server hours***		Total Hours
3:00–4:00			3		3
4:00–5:00	20	40	3		3
5:00–6:00	40	68	3	4	7
6:00–7:00	56	88	3	4	7
7:00–8:00	64	93	3	4	7
8:00–9:00	58	85	3	4	7
9:00–10:00	54	64	2	3	5
10:00–11:00	20	20	2	2	4
11:00–12:00				2	2
Total	312				
# of hours			22	23	45
Wage per hour					$ 5.50
Total daily wage					$ 247.50

Table 8-04(B1) (Continued)

Tuesday Projection						
Hours	Cust.*	In hall**	Server hours***		Total Hours	
3:00–4:00			2		2	
4:00–5:00	10	17	2		2	
5:00–6:00	15	29	2	2	4	
6:00–7:00	28	43	2	3	5	
7:00–8:00	30	41	2	3	5	
8:00–9:00	23	27	2	2	4	
9:00–10:00	8	8		2	2	
10:00–11:00				2	2	
Total	114					
# of hours			12	14	26	
Wage per hour					$	5.50
Total daily wage					$	143.00

Saturday Projection						
Hours	Cust.*	In hall**	Server hours***		Total Hours	
3:00–4:00			4		4	
4:00–5:00	20	40	4		4	
5:00–6:00	40	68	4	3	7	
6:00–7:00	56	88	4	4	8	
7:00–8:00	64	93	4	4	8	
8:00–9:00	58	85	4	4	8	
9:00–10:00	54	64	4	3	7	
10:00–11:00	20	20		3	3	
11:00–12:00				3	3	
Total	312					
# of hours			28	24	52	
Wage per hour					$	5.50
Total daily wage					$	286.00

Wednesday Projection						
Hours	Cust.*	In hall**	Server hours***		Total Hours	
3:00–4:00			2		2	
4:00–5:00	10	22	2		2	
5:00–6:00	25	42	2	3	5	
6:00–7:00	35	55	2	3	5	
7:00–8:00	40	52	2	3	5	
8:00–9:00	25	32	2	3	5	
9:00–10:00	15	15		3	3	
10:00–11:00				2	2	
Total	150					
# of hours			12	17	29	
Wage per hour					$	5.50
Total daily wage					$	159.50

Sunday Projection						
Hours	Cust.*	In hall**	Server hours***		Total Hours	
2:00–3:00			1		1	
3:00–4:00	5	9	1		1	
4:00–5:00	8	18	1	2	3	
5:00–6:00	20	29	1	2	3	
6:00–7:00	18	25	1	2	3	
7:00–8:00	14	19	1	2	3	
8:00–9:00	10	12		2	2	
9:00–10:00	4	4		2	2	
10:00–11:00				2	2	
Total	79					
# of hours			6	14	20	
Wage per hour					$	5.50
Total daily wage					$	110.00

Table 8-04(B1) (Continued)

Thursday Projection						
Hours	**Cust.***	**In hall****	**Server hours*****		**Total Hours**	
3:00 – 4:00			2		2	
4:00 – 5:00	18	28	2		2	
5:00 – 6:00	20	45	2	3	5	
6:00 – 7:00	50	77	2	3	5	
7:00 – 8:00	55	75	2	3	5	
8:00 – 9:00	40	47	2	3	5	
9:00 – 10:00	15	15		2	2	
10:00 – 11:00				2	2	
Total	198					
# of hours			12	16	28	
Wage per hour					$	5.50
Total daily wage					$	154.00

* The customer counts are the ones that leave by the end of each hour.

** In-hall customer count of each hour is estimated with a half of the next hour customer added to the current ones. For example, during the hour of 6-7, 17 customers are finishing their meal, and 11 additional customers are being served.

*** Server hours stand for the number of servers working in each hour. For example, from 5 to 6 o'clock, 4 servers are scheudled to work to serve 28 customers in the dining hall. During this one hour, 4 hours' payroll is being incurred.

More details are introduced in the text.

Table 8-04(B2) Weekly Wage Total – Front – Y7

Day	Hour	Wage	Amount
Monday	25	$ 5.50	$ 138
Tuesday	26	$ 5.50	$ 143
Wednesday	29	$ 5.50	$ 160
Thursday	28	$ 5.50	$ 154
Friday	45	$ 5.50	$ 248
Saturday	52	$ 5.50	$ 286
Sunday	20	$ 5.50	$ 110
Weekly Total	225		$ 1,238
Annual Total (52 wks)			$ 64,350

* These are tipped employees.

Table 8-04(C) Total Labor Cost breakdown (min vs. max)

	Y-2 (min)	Y-7 (max)
Salaries	$ 150,000	$ 150,000
Wage – Front	$ 60,918	$ 64,350
Wage – Back	$ 144,872	$ 144,872
Contingency Labor	$ 50,000	$ 100,000
Net Labor Costs	$ 405,790	$ 459,222
Labor related	$ 101,605	$ 101,605
Total Labor Costs	$ 507,395	$ 560,827
Total Revenues Projected	$ 1,544,215	$ 1,960,400
Labor cost %	32.9%	28.6%

Table 8-05 Long-term Projection of the Next 7 years

The Pro Forma Income Statements of the next 7 years R&B Grill	FC	VC/cust	Y-2	V.A.%	Y-3	Y-4	Y-5	Y-6	Y-7	V.A. %
	Customer #		58,916		60,683	62,504	64,379	66,310	67,600	
	Avg. Check*		$26.35		$26.85	$27.35	$27.90	$28.45	$29.00	
	Increase % (exp. Inflation)				1.9%	1.9%	2.0%	2.0%	1.9%	
Revenues			1,552,437	100.0%	1,629,351	1,709,484	1,796,177	1,886,533	1,960,400	100.0%
Cost of sales	Cost %	33.5%	520,066	33.5%	545,833	572,677	601,719	631,989	656,734	33.5%
Gross Profit			1,032,370	66.5%	1,083,519	1,136,807	1,194,458	1,254,544	1,303,666	66.5%
Operating Expenses	FC (annual)	VC/cust								
Labor expenses	per labor schedule		507,395	32.7%	517,543	527,894	538,452	549,221	560,827	28.6%
Restaurant operating expense		$ 1.25	73,645	4.7%	75,854	78,130	80,474	82,888	84,500	4.3%
Marketing expenses	$ 20,000		20,000	1.3%	20,000	20,000	20,000	20,000	20,000	1.0%
Energy (Utility) expenses	$ 11,652	$ 0.41	35,808	2.3%	36,532	37,279	38,047	38,839	39,368	2.0%
Administrative expenses	$ 58,980	$ 0.72	101,400	6.5%	102,672	103,983	105,333	106,724	107,652	5.5%
Repair & maintenance expenses		$ 0.30	17,675	1.1%	18,205	18,751	19,314	19,893	20,280	1.0%
Rent expenses	$180,000	$ –	180,000	11.6%	180,000	180,000	180,000	180,000	180,000	9.2%
Depreciation – FF&E	$ 24,000	$ –	24,000	1.5%	24,000	24,000	24,000	24,000	24,000	1.2%
Depreciation – Leasehold improve.	$ 24,000	$ –	24,000	1.5%	24,000	24,000	24,000	24,000	24,000	1.2%

Depreciation – Smallware	$ 24,000	$ –	24,000	1.5%	24,000	24,000	24,000	24,000	24,000	1.2%
Total Operating expenses			1,007,922	64.9%	1,022,807	1,038,036	1,053,620	1,069,565	1,084,627	55.3%
Operating profits			24,448	1.6%	60,712	98,770	140,838	184,980	219,039	11.2%
Interest expense (income)			18,000	1.2%	18,000	18,000	18,000	18,000	18,000	0.9%
Income before income taxes			6,448	0.4%	42,712	80,770	122,838	166,980	201,039	10.3%
Income taxes (35%)			2,257		14,949	28,270	42,993	58,443	70,364	
Net Income (loss)			4,191	0.3%	27,763	52,501	79,845	108,537	130,675	6.7%

Balance Sheet records – projection – the R&B Grill's financial positions will be maintained at the current values.

Average Assets	1,135,705
Average Liabilities	421,000
Average Equity	714,705

Projected Ratio Summary	Y-2	Y-3	Y-4	Y-5	Y-6	Y-7
Assset Turnover	1.37	1.43	1.51	1.58	1.66	1.73
Return on Sales	0.27%	1.70%	3.07%	4.45%	5.75%	6.67%
Return on Assets	0.37%	2.44%	4.62%	7.03%	9.56%	11.51%
Return on Equity	0.59%	3.88%	7.35%	11.17%	15.19%	18.28%

Y-2; Assets at $1,135,705, with the liabilities of $421,000 and the owners' equity at $714,705. This assumption of the financial position is made only because the balance sheets cannot be easily projected. If R&B Grill progresses as projected, its financial position will be much better to the owners than what is shown in the projection.

8.6 ANALYSIS OF THE PROJECTED PERFORMANCE

According to the projection, R&B Grill's Asset Turnover will improve from 1.37 in Y-2 gradually to 1.73 in the peak year (Y-7). Its Return on Sales (ROS) will also gradually increase from 0.27% in Y-2 to close to 6.7% in the seventh year. The ultimate ROS is close, if not above, to those of major national chain competitors. For an independent local business, it may be better to maintain ROS a little lower with prices a little high to attract customers' attention. This may sound unusual, but lower profits from higher prices mean that the business spends more on the quality of products and services.

With the projected operational results, its Return on Assets (ROA) grows from the 0.37% to over 11.5% and so does its Return on Equity (ROE) from 0.159 to over 18%. In conclusion, the operating forecast of the next several years until the business matures is quite positive and the investors and the management can feel comfortable to pursue their plan.

Not only this projection is prepared within the reasonably achievable range, but also is it prepared quite conservatively. In other words, this plan is not overly optimistic. What is not included in the projection process is the analysis of the business environments like the customer behavior trends, competitors, studies about necessary supply chain, and potential regulatory pressure. These are the topics of Strategic Management. Restaurateurs must continuously investigate these factors and prepare their business strategies and tactics to navigate their operations through the changes of them.

8.7 CONCLUSION

This chapter introduced a method of conducting long-term projection. It used existing operational data of Y-1 as the base. Revenues projection was made accommodating two conditions. One is the physical size of the property with 100 seats and maximum seat turnover that can be achieved. The other is the reasonable average check price a customer would be willing to pay. In this approach, naïve forecasting method and the probabilistic method have been used to forecast conservative results as possible.

The Costs of Sales were projected at 33.5% of the revenues of each year. This was slightly overestimated to accommodate potential fluctuation of individual years' unique situation of merchandise supplies. Labor Costs projection included the wages of front-of-the-house employees and of back-of-the-house employees. To these, salaries were added along with payroll-related expenses, and contingency labor funds.

All other expenses were projected using the forecasted customer counts to determine the amount of total variable cost of each expense and the amount of fixed cost was added. The amounts of fixed cost and unit variable cost (per customer) have been obtained through applying the Regression Analysis technique.

Finally, Table 8.6 was presented to show the result of the long-term projection. In this scenario, R&B Grill is expected to reach its peak in the seventh year from the opening. The next phase is to determine the market value of R&B Grill as of now to verify whether this business is good enough for the investors (Equity holder: Rachel and Brad, and Jane). The next two chapters introduce conceptual and technical details of this process known as "Valuation."

Table 8-06 Long-term Projection of the Next 7 years with Cash Flows

The Pro Forma Income Statements of the next 7 years R&B Grill								
	FC	VC/cust	Y-2	Y-3	Y-4	Y-5	Y-6	Y-7
	Customer #		58,916	60,683	62,504	64,379	66,310	67,600
	Avg. Check*		$26.35	$26.85	$27.35	$27.90	$28.45	$29.00
	Increase % (exp. Inflaticn)			1.9%	1.9%	2.0%	2.0%	1.9%
Revenues			1,552,437	1,629,351	1,709,484	1,796,177	1,886,533	1,960,400
Cost of sales	Cost %	33.5%	520,066	545,833	572,677	601,719	631,989	656,734
Gross Profit			1,032,370	1,083,519	1,136,807	1,194,458	1,254,544	1,303,666
Operating Expenses	FC (annucl)	VC/cust						
Labor expenses	per labor schedule		507,395	517,543	527,894	538,452	549,221	560,827
Restaurant operating expense		$ 1.25	73,645	75,854	78,130	80,474	82,888	84,500
Marketing expenses	$ 20,000		20,000	20,000	20,000	20,000	20,000	20,000
Energy (Utility) expenses	$ 11,652	$ 0.41	35,808	36,532	37,279	38,047	38,839	39,368
Administrative expenses	$ 58,980	$ 0.72	101,400	102,672	103,983	105,333	106,724	107,652
Repair & maintenance expenses		$ 0.30	17,675	18,205	18,751	19,314	19,893	20,280
Rent expenses	$ 180,000	$ –	180,000	180,000	180,000	180,000	180,000	180,000
Depreciation – FF&E	$ 24,000	$ –	24,000	24,000	24,000	24,000	24,000	24,000
Depreciation – Leasehold improve.	$ 24,000	$ –	24,000	24,000	24,000	24,000	24,000	24,000

Depreciation – Smallware		$ 24,000	$ –	24,000	24,000	24,000	24,000	24,000	24,000
Total Operating expenses				1,007,922	1,022,807	1,038,036	1,053,620	1,069,565	1,084,627
Operating profits				24,448	60,712	98,770	140,838	184,980	219,039
Interest expense (income)				18,000	18,000	18,000	18,000	18,000	18,000
Income before income taxes				6,448	42,712	80,770	122,838	166,980	201,039
Income taxes (35%)				2,257	14,949	28,270	42,993	58,443	70,364
Net Income (loss)				4,191	27,763	52,501	79,845	108,537	130,675
Cash Flows Projection of the next 7 years									
Cash Flows from Operating Activities	Net Income			4,191	27,763	52,501	79,845	108,537	130,675
	plus Non-cash expenses			72,000	72,000	72,000	72,000	72,000	72,000
Total CF from Operating Activities				76,191	99,763	124,501	151,845	180,537	202,675
Cash Flows from Investing Activities	Estimated additional inv.			–	15,000	20,000	25,000	30,000	40,000
Free Cash Flows				76,191	84,763	104,501	126,845	150,537	162,675
Cash Flows from Financing Activities	@ 7.5% interest rate			71,876	71,876	71,876	71,876	71,876	71,876
Net Cash Flows				4,315	12,887	32,625	54,969	78,661	90,799

CHAPTER 9

RISK, RETURN, AND THE TIME VALUE OF MONEY (TVM)

9.1 INTRODUCTION

All investments are the actions of putting currently available resources into an opportunity expecting better returns in the future. However, expected returns are not guaranteed. All investments always involve uncertainty that actual returns may and will be different from the expected outcome. The uncertainty that causes the difference between the expected return and the actual outcome is considered the risk involved in investments. This chapter introduces the concept and the relationship of *Risk* and *Return* that are the inherent conditions of all investing activities.

9.2 VALUE-ADDING THROUGH THE SERVICE EXCHANGE PROCESS

Figure 9.1 displays the mechanism of a business that adds value to the equity holders' investments. A business is created in the beginning by financing its necessary capital to prepare its assets from two sources that are creditors and equity holders. This concept is widely known as the fundamental accounting equation of "$A = L + E$" as illustrated in the second column from the left. Once the necessary assets are formed, the business conducts its operations to generate revenues. Conducting operations naturally uses the assets. The effectiveness of this is presented

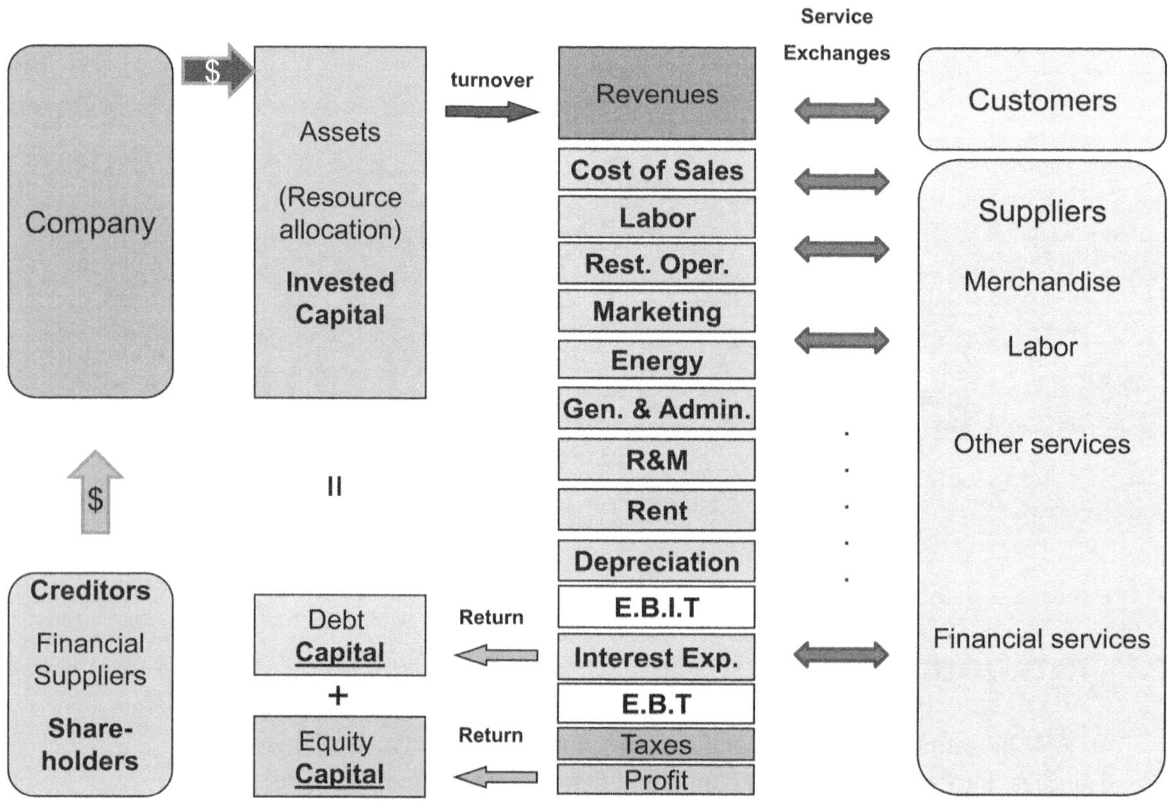

Figure 9.1 Value-adding through the service exchange process.

in the asset turnover ratio. The third column shows the structure of the Income Statement to illustrate necessary activities in operation that incur expenses to support the activity of generating revenues. As shown in the diagram, operating expenses are deducted from the revenues to estimate the operating profit. Operating expenses are necessary spending to compensate individual service providers. The only exception is the depreciation expense, which will be explained further in the later section. The service providers in operations include the suppliers of the necessary material, labor, and other utilities and services, to name only a few. This process requires mutual satisfaction between the payer and the receiver. The small arrows under the title of "Service Exchanges" indicate this relationship. First, customers must be satisfied with the products and services they receive by paying the price. What is also important, however, is that all other stakeholders that support the business to serve customers – such as employees, utility service providers, and even the landlord, to name a few – must be satisfied with the payment they receive by providing their services. Otherwise, the business itself may not be sustained for long. After taking out these operating expenses, the business reports the leftover from the revenues as its operating income that is presented in Figure 9.1 as Earnings before Interest and Taxes or EBIT.

From the operating income (EBIT), the business pays interests to the creditors. It is the return on creditors' investments. Once the interest is taken out, the remaining amount is reported as the taxable income. It is usually presented as the earnings before tax (EBT). It is important to notice that all expenses taken out up to this point (before EBT) are tax deductible. The more the business spends on these expenses, the less amount of tax it will pay. The advantage of spending more on individual operating expenses is that a business may achieve better customer loyalty, higher employee morale, providing better working conditions, to name a few. These will eventually bring about better results by providing added value to individual stakeholders of the business. Besides, as an additional advantage, the business saves on its income tax expense, too.

After the income tax is deducted from the EBT, the net income is finally presented and it is returned to the equity investors. This is how value is added to the investments made by the owners through service exchanges. It is only accomplished when the equity investors receive their return at the last step after the completion of entire business transactions at each step. Revenue generating activities collect fees from customers by serving them with products. The Cost of Sales is the amount of the merchandise directly spent for revenues. It serves merchandise suppliers and customers as well. Labor expense is the compensation provided to the employees who have engaged in the service activities. Like these, there are continuous transactions between the firm and various types of service providers in addition to customers. It is not only the customers who should be satisfied with the products and services they consume. All other service providers, including the creditors, must also be satisfied with the compensation they receive for their contribution. This is how their values are added. Only when all these stakeholders are fully compensated, the equity holders will finally receive their returns. This is the brief illustration of how business activities are summarized in its accounting reports as presented in Figure 9.1.

9.3 THE SOURCE OF UNCERTAINTY

The value-adding process described in the previous section starts from the investments made by the owners and creditors and ends with providing expected returns to the owners. Conducting this process to achieve the ultimate goal of providing returns to the owners is the responsibility of managers. When this process is viewed from a long-term futuristic perspective, it is easy to notice that managing a business must deal with a serious magnitude of uncertainty for every step of service-exchange process. This section introduces the causes of uncertainty that imposes risk on the business. Figure 9.2 "Sources of Uncertainty" illustrates the source and direction of uncertainty.

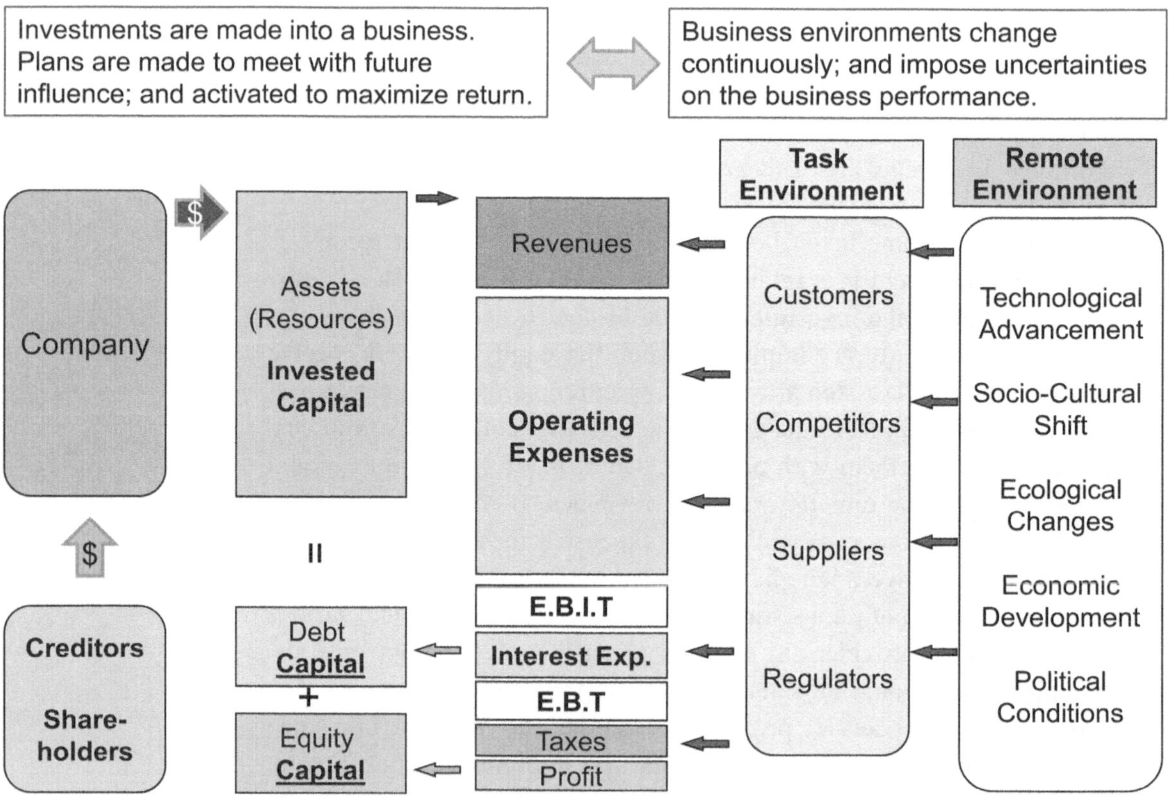

Figure 9.2 Sources of uncertainty.

All businesses are surrounded by influential conditions that affect the performance and the result. Those conditions are called the environmental factors of business and they create uncertainty. R&B Grill, even as a small private restaurant company, has to face and deal with the changes of its environmental factors. In Figure 9.2, the rounded rectangle under the title of "Task Environment" includes four major components of Customers, Competitors, Suppliers, and Regulators. They represent the environmental factors that are considered to influence the tasks of business firms directly.

R&B Grill must observe the changes of its customers' behavior patterns in their consumption habits and adjust its strategies to accommodate them. There may not be immediate changes in the customer behavior patterns but long-term projections of this must be conducted to prepare for gradual adjustments in the future. One of the important elements of this includes the shift of the demographic attributes of the potential customer base in the region. As time goes by, R&B Grill must face its customers growing older in the region. If its current target customers belong to the population between the age of 30 and 50, they will become 40 to 60, in about 10 years. It may be difficult to attract them by that time with the same products and services. R&B Grill must prepare its own plans to adjust its concepts to sustain its growth by following through the changes of its customer base. Another factor that creates uncertainty is the activities of competitors – direct or indirect. Supply chains and regulations are also important factors to consider in assessing future uncertainty. To be more specific, supply chains must be addressed not only from the sides of the merchandise but also from those of the labor and financial sides. These surrounding factors interact with one another constantly and create future uncertainty. The managers of a business must keep their eyes open for any possible changes of these factors and remain prepared for future conditions.

The next rounded rectangle under the title of "Remote Environment" includes five factors that interact with one another and create long-term effects on business performance. Those long-term effects are usually called "trends." These conceptual factors of this environment interact with other factors and create huge impacts on the Task Environmental factors. R&B Grill, for example, must be able to observe the remote trends that will influence its business in the long run and be prepared to sustain its business in the future to continuously satisfy its owners who are its ultimate stakeholder group.

9.4 EXAMPLES OF ENVIRONMENTAL ANALYSIS

One good example of this can be introduced with the crude oil price. Several years ago when crude oil price was rising sharply in the international market, the Southwest Airline Company had predicted it correctly and had signed a contract before the actual price went higher with its gas suppliers to purchase fuel at a lower price. As the result of the contract, Southwest Airline Company enjoyed huge operating profit when all the other airline companies were suffering high fuel cost for their operations. The main reason of the crude oil price increase at that time was the large volume of oil imported by China. During those days, China was expanding its economy rapidly. Many of the airline companies that suffered high fuel cost followed the example of Southwest Airline, but the crude oil price declined a lot because China's import of crude oil did not last long as many people had predicted. Consequently, all these companies who just blindly imitated Southwest Airline had to suffer hundreds of million dollars of operating losses due to their contracts of purchasing fuel at a higher price when the market price was dropping. This is an example of a business firm that took great advantage of its prediction of the ecological factor mixed with the economic and political factors developed in the Remote Environment.

Another example can be introduced with Panera Bread Company, an up-scaled limited service restaurant chain. Panera Bread Company was a successful firm with its unique service concept. However, its concept became extremely popular after the Great Recession of 2009. Before the Great Recession, when the entire economy was rapidly expanding, many chain restaurant companies invested heavily into high-end service concepts. Most of these expansions were financed through issuing long-term debt taking advantage of the governmental policy of lower interest rate. This was an example of economic expansion driven by political factors. During the same time, housing market was mushrooming due to the same reason. However, when this overheated economy was chilled almost overnight, consumer market shrank, and the restaurant businesses that expanded rapidly could not sustain their operations. When customers turned to lower priced, limited-service concepts, Panera Bread Company's operating profits were skyrocketing. As the result, the stock price, which is the official proxy of the firm's market value, jumped and leaped when many of its high-end competitors were suffering huge financial distress. This is a successful example of the company's healthy policy that met with the mix of economic factors that drove customers to its stores.

In summary, the shift of business environmental factors creates uncertainty and it imposes risk on every business. It is the responsibility of managers to watch closely the changes of their business environments and to prepare for the future to sustain the expected outcome. The next section introduces how to convert these conceptual approaches to a numerical one.

9.5 EVERYTHING FUNNELS DOWN TO THE FINANCIAL RISK OF A BUSINESS

As explained in the previous section, the shift of external environmental factors creates uncertainty in the future path of a business and the future result of a business may be different from the expected outcome. The difference between the actual result and the expected outcome is the definition of a risk. In this context, risk of a business performance can also be described as the variance between the expected outcome and the actual result.

In business, results are always presented in financial terms, and the ultimate result in financial term – either expected or actual – is always the rate of return provided to the owners. As presented in Figure 9.1, there are multiple steps to reach the final result from serving customers in the operating process. Because every step in the operating process incurs costs, the variance between the expected costs and actual ones also become the source of risks. These are considered as the internal risk. Although internal risks are created by the changes of the external factors, they are usually transformed and presented as financial distress due to the cost structure of a business. A larger amount of Fixed Costs requires higher amount of revenues to break even. This will increase the pressure to the operators. In such a case, however, once the operations surpass the break-even point, the profit margin will increase rapidly. On the other hand, the larger portion of variable costs will require lower amount of revenues to break even, which will result in the lower pressure to the operators. However, a larger mix of variable costs will lower the profit margin even at a higher amount of revenues. It is the responsibility of the management to determine the most reasonable cost structure in its operating system. This will lead to the next phase of the risk involved in operations caused by the financial structure.

9.6 OPERATING RISK IMPOSED BY THE DEBT CAPITAL

A business must provide expected returns to satisfy the investors and the return is generated only through operations. In this context, operating is the only mechanism that generates financial returns. To operate effectively, operators must have necessary supplies of assets. Hence, a firm must raise necessary funds. This process is known as financing activities of a firm. Financing activities take two different forms. One is the debt, and the other is the direct investment made by the owners. Both groups are investing with their demand of their returns. The returns demanded by the creditors are the interest expense to the firm. Since the return of interest is guaranteed along with the payment of the principal, the financial risk of the creditors is lower than that of the owners. For this reason, the interest rates become eventually lower than the rate of the return required by the owners. Consequently, a business that uses borrowed funds can lower its burden of providing returns. Another benefit of using debt capital is that the interest is also a tax-deductible expense. By taking out its interests from the operating profits (EBIT), the business may save in its income tax expenses. This is a practical advantage to a business and restaurant businesses are no exception. One caution is, however, the interest is also an expense that must be paid in cash. The operations must be strong enough to generate EBIT to cover the interest. In addition to this, the business must generate enough cash flows from its operations to pay the principal amount of its debt.

In the previous chapter, R&B Grill's annual debt service amount was introduced as $71,816. This amount has been estimated by applying the interest rate of 7.5% to the total long-term debt of $421,000 for the term of 8 years. The calculation can be easily conducted using a spreadsheet program such as Excel. Today's Internet search engines (such as Google) also provide plenty of support for technical needs like this. Table 9.1 shows how R&B Grill can pay off its entire debt of $421,000 by making the annual payment of $71,816 for 8 years.

Table 9-01 R&B Grill – Loan Amortization Table

Loan amount	$ 421,000	Interest Rate	7.50%	Term	8
Payment	$ 71,876				

Term (years)	Payment (Debt Service)	Rate	Interest $	Principal	Balance
0					$ 421,000
1	$ 71,876	7.50%	$ 31,575	$ 40,301	$ 380,699
2	$ 71,876	7.50%	$ 28,552	$ 43,324	$ 337,375
3	$ 71,876	7.50%	$ 25,303	$ 46,573	$ 290,802
4	$ 71,876	7.50%	$ 21,810	$ 50,066	$ 240,736
5	$ 71,876	7.50%	$ 18,055	$ 53,821	$ 186,916
6	$ 71,876	7.50%	$ 14,019	$ 57,857	$ 129,058
7	$ 71,876	7.50%	$ 9,679	$ 62,197	$ 66,861
8	$ 71,876	7.50%	$ 5,015	$ 66,861	$ –
Total	$ 575,008.61		$ 154,009	$ 421,000	

The table shows the long-term debt of $421,000 is gradually diminishing finally to zero ($0) in 8 years by making the annual debt-service payment of $71,876. The amount of annual debt-service includes annual interests in it. Once the annual interest is taken out, the remaining amount is used to pay the debt itself (principal). It must be noticed that the annual interest amount shrinks each year as the balance of the debt declines although the interest rate remains the same. As the shrinking amount of interest is taken out from the constant amount of annual debt-service, the principal payment is increasing each year.

This mechanism introduces the relationship between the risk and the return related to the passage of time as explained in the following. The creditors have taken the risk by giving loans to R&B Grill. By doing so, they demand their return of 7.5% per year. The creditors' risk is the highest in the beginning when the unpaid principal amount is the largest. Hence, they are taking the largest amount of return from the annual debt services paid by R&B Grill. As the balance amount gradually decreases, the interest amounts also decline because the creditors' risk is reduced.

However, the borrower must generate the same amount of cash from its operations to handle this burden each year until its debt is completely paid out. As pointed out earlier, if the business takes an excessive amount of debt beyond its operational capability of generating enough cash to handle the debt-service amount, the business will face serious financial distress. R&B Grill, in a conservative projection, is supposed to generate enough amounts of Free Cash Flows even during the earlier years to handle its annual debt-services at the interest rate of 7.5%. Even if the interest rate goes up to 10.0%, the annual debt-service amount will be about $79,000. This will be smaller than the expected Free Cash Flows of the third year. The first couple of years' debt-services, which exceeds the amount of the Free Cash Flows, can be handled with the reserved cash ($640,225 as presented in Table 7-02) in the current balance sheet. Restaurateurs must maintain reasonable financial flexibility by keeping safety nets in financial plans to cover unexpected situations.

9.7 COMPENSATION OF THE EQUITY INVESTORS' RISK

The previous section explained the risk taken by the creditors, and the financial burden to the borrowing firm. This section introduces how equity investors determine their expected return, which imposes the ultimate financial burden to the firm that uses their invested funds.

Equity investors invest their wealth to a business with no guarantee on their return. For this reason, they demand higher return to compensate their risk. To separate the "investors" from the "creditors," they will be referred to as "equity holders" or "owners" from now.

There is a widely used method to estimate the reasonable rate of return on the investments made by the equity holders. Different from creditors, owners have to consider multiple factors for their investment decision. Because their returns are completely dependent upon the future performance of the business to which they invest, they start building up their demand on a solid minimum return that is guaranteed. This guaranteed return is considered as risk-free rate, and the yield rate of Treasury Bonds is frequently used as the proxy of this. Of many types of the Treasury Bonds, for a fair comparison with the investment into a business firm, the yield rate of the 10-year Treasury Bonds is often used as a benchmark because investing into common stocks usually takes long-term plans for real returns. As the yield rate of Treasury Bonds is very low due to its nature of being risk-free, equity investors demand higher rate than this by adding other elements for their required return on investment. The following explains other elements added to the risk-free rate.

9.7.1 Risk Involved in the Entire Market

Compared with Treasury Bonds, business firms in the private sectors carry much higher risks. For this reason, equity investors want to compensate the risk by demanding extra return on their investment beyond the risk-free rate. The demand of extra return is usually described as "Normal Market Premium" or "Risk Premium." Differently put, equity investors want to have "Premium" return from private business firms to compensate their higher risk by giving up the guaranteed risk-free rate from the Federal Government.

There is no fixed rule to determine this premium in numerical terms, but many experts use the historical average annual return from the stock market above the yield rate of 10-year Treasury Bonds. According to the records provided by Damodaran[1], the market return from the S&P 500 during the past 50 years is 9.61%. During the same period, the 10-year Treasury bond has yielded 6.71% return. From these data, the market premium (or risk premium) of choosing the stock market for investments over the Treasury Bonds is 2.90%. In other words, the S&P 500 has provided 2.9% higher returns than 10-year Treasury Bonds during the past 50 years. During the past 10 years from 2006 to 2015, the return from the S&P 500 is 7.25% and the return from the 10-year Treasury bond is 4.71%. The risk premium during this period is 2.53%.

Risk premium is the additional return demanded by investors to compensate their risk beyond the risk-free rate that is guaranteed if they invest into the Treasury bills or bonds. Although many experts have provided historical data as benchmark risk premiums as introduced in this section, the real risk premium for individual investors can be considered quite subjective values depending on their demands. Some will be satisfied with around 3% higher return than the risk-free rate but some may

[1] http://pages.stern.nyu.edu/~adamodar/New_Home_Page/datafile/histretSP.html.

want much more return to compensate the risk they take by choosing the market. The structure of higher demand can be summarized as follows:

Expected Return = Risk-free rate + Normal Market (or Risk) Premium

It is important to remember that the risk premium mentioned so far is about the entire market rather than an individual firm. As the level of risk varies from one firm to another, we need to estimate the additional risk involved in individual businesses and the appropriate rate of returns to compensate the risk. The next section explains this aspect.

9.7.2 Risk Involved in an Individual Firm

The business that uses investors' financial contributions must provide the highest return that will satisfy them. Individual businesses carry different levels of risk because their market conditions are unique and their strategies are different. For publicly traded corporations, there is an easy index that represents their risk level. It is called Beta. Beta is publicly available information that indicates the risk level of a public company. It is a simple ratio that reflects the company's stock price fluctuation over the entire market fluctuation. If the beta of a company is 1, it means the risk of the business is considered the same as that of the entire market. If it is larger than 1, the company is considered riskier than the market. If the beta is less than 1, the company is considered safer or more stable than the market for investing.

The beta value of a public company is used to determine the required rate of return of investors by multiplying it to the estimated risk premium. It is important to reiterate that investors also have set their non-negotiable risk-free rate before asking for their premium. Thus, the final equation that represents the required rate of return finally demanded by investors (equity holders) of a firm looks like the following:

Required rate of return = Risk-free rate + Beta × Risk Premium

The equation introduced above is called the Capital Asset Pricing Model (CAPM). Although it has been challenged often for its weaknesses, it is still considered the standard method of estimating the most reasonable rate of return for equity investing and is widely used. It must be noticed that two of the three elements – risk-free rate and the Beta of a company – are publicly available. The value of the Risk Premium, on the other hand, is a subjective term. Individual investors may use their own decisions or they may use publicly available information as introduced above.

9.7.3 Application of the CAPM to R&B GRILL

Let's apply all these to R&B Grill. Its investors can invest into the 10-year Treasury Bond that yields about 1.7% per year at the time of writing this script, with no fear of losing. This is their risk-free rate guaranteed by the US Federal Government. However, they have chosen a small private business that must carry much higher risk. By choosing a risky opportunity in the commercial sector, they must demand higher return that will compensate the risk. Let's assume they are following the general guidelines provided in the previous section.

By following such public guideline, they apply at least 3% return per year above the risk-free rate (1.7%) annually. This (3% higher demand) is the Normal Market Premium (or Risk Premium) they want additionally when choosing a commercial sector for their investment. This Risk Premium of 3% is also slightly higher than the 2.53% per year that was introduced in the previous section. The last step is to assess the risk of R&B Grill itself. R&B Grill, as a small restaurant business, must carry a higher level of risk. The restaurant industry is traditionally well known for its high risk due to its volatilities in revenues and profits over time. Since R&B Grill is a private company, its beta value is not available publicly. However, it is possible to use the beta of similar public companies that share common attributes in their concepts. The major competitors that are similar to R&B Grill in the clientele and concepts with compatible services at similar price range include the Darden Restaurants, Outback Steakhouse, Chilli's, and their counterparts. The beta values of those companies can be used as the proxy with reasonable adjustments.

Quick Internet search of several public restaurant chains reveal that their beta values range from as low as 0.26 of the Darden Restaurant to as high as 1.2 of the B.J. Restaurant Inc. If the investors think the risk of R&B Grill is compatible to that of the B.J. Restaurant Inc., the riskiest company in the list, they can use 1.2 for the beta, and the CAPM can be applied to determine their reasonable return on investment.

$$\text{Required rate of return} = \text{Risk-free rate} + \text{Beta} \times \text{Risk Premium}$$

$$\begin{aligned}\text{Investors' ROI} &= 1.7\% + 1.2 \times 3\% \\ &= 5.3\%\end{aligned}$$

The result of the 5.3% shown above is the rate of return demanded by the investors of R&B Grill when its risk level is considered the same as that of B.J. Restaurant Group. It is above the risk-free rate of 1.7% by 3.6%. This additional return of 3.6% is the multiple of the 3% risk premium and the specific risk of R&B Grill itself estimated by the investors by substituting the Beta (1.2) of the B.J. Restaurant Inc. By multiplying the risk premium to the risk index of the business (beta), investors determine the rate of return they demand from the business they have chosen for their investments. Then, the result is added to the risk-free rate to determine the final rate of return on their investments. Finally, the equity investors want to have return of 5.3% on their investments into R&B Grill.

So far, the required rates of returns of the two groups of investors – creditors and the owners – have been introduced in relation to their perceived risks. In the case of R&B Grill, the creditors want to have at least 7.5% of return, while the equity investors want to have 5.3% annual return on their investments. R&B Grill must generate enough profits through its operations to provide these returns. To R&B Grill, these are called its cost of debt capital and the cost of equity capital, respectively. Financial suppliers' demands become the firm's costs.

In conclusion, interest rates on long-term debt and the required ROI by the equity investors represent the level of risk perceived by each group of the business that uses their funds. Differently put, financial suppliers impose their perceived risks on the business and demand returns that will compensate them. In the case of R&B Grill, its creditors demand 7.5% return on their investments, while its equity investors want at least 5.3% return on their investments per year because they think those are the reasonable rates that compensates the risk they have taken. The next section introduces how these demands of returns are applied at the time of investing.

9.7.4 Time Value of Money with Expected Return

Almost all books on the topic of financial management introduce a lot of formulas that are designed to solve highly complicated problems related to the concept of the Present Value (PV) and the Future Value (FV) of money. They usually deliver a large amount of information with various applications to explain the concept of the Time Value of Money (TVM) on many occasions. This book, in contrast, presents a simplified application of the concept of the TVM using the example of R&B Grill.

The creditors and the equity investors – that are financial suppliers – of R&B Grill invested expecting returns in the future that will compensate their risk. This statement inherently contains the concept of time. They invest in present amount expecting to have returns in future amount. The return they expect to receive must be larger in amount to compensate the risk taken at the time of investing. In this context, the investment is in PV and the return is in FV and the PV (which is investment) must grow over time into a larger amount in the FV and that should be returned to the investor. This relationship is presented in an equation that follows:

$$\text{Future Value} = \text{Present Value} \times (1 + \text{growth rate})^t$$

It must be noticed that the growth is exponential. The following example explains the effect of exponential growth and the growing value of the investment over time.

Example: *If you are investing \$1,000 now with the expected growth of 10% return per year, your investment will grow into \$1,100 at the end of the first year. The following equation explains this numerically.*

$$\text{Future Value} = \text{Present Value} \times (1 + \text{growth rate})^t$$

$$\$1{,}100 = \$1{,}000 \times (1 + 0.1)^1$$

In this scenario, the \$1,000 that you are investing now is the PV and the expected amount of \$1,100 at the end of the first year is the FV. Although the amount grows in a year by 10%, it is also the risk you have taken. As such, the amount of growth at 10% just compensates your risk and the two different amounts in time – \$1,000 now and \$1,100 in one year from today – are the same in value when the risk is compensated.

The example introduced above indicates that only when your investment grows as expected (10% per year), the FV is the same as the PV despite the two different amounts in time. In other words, if the actual growth were only 9% instead of 10%, the FV does not satisfy your expectation. Thus, the value of the return (which is 9%) is smaller than the value you were expecting at the time of investing. This is the main point of TVM introduced in this book, which departs from many other books by providing more practical views of investments and expected return. The value of future return, regardless of its amount, remains the same only when the investor's expectation is fully compensated.

The interpretation provided above indicates that invested amount must grow at the expected rate over time into larger amounts. The longer it takes, the larger it must become. If higher growth rates are expected, the larger it has to grow. These two elements – time and growth rates – are the risk taken by investors at the time of investing.

Once financial suppliers invest into a business firm, the firm must provide returns continuously as expected by each group of financial suppliers. Differently put, investors' expectation becomes the firm's risk. Let's clarify this relationship of the investors and the firm using the example of R&B Grill's long-term debt.

The TVM formula introduced above was about the growth of a PV into a FV. The same formula can be reversed to calculate the PV of an expected FV as follows:

$$\text{Present Value} = \text{Future Value } (1 + \text{growth rate})^{t}$$

This formula can be presented as follows:

$$\text{Present Value} = \text{Future Value } (1 \quad (1 + \text{growth rate})^{t})$$

Either of these formulas can be used to determine the PV of any FV expected. When this approach is used, FV is "discounted" into PV. The algebraic element of $(1 \quad (1 + \text{growth rate})^{t})$ used in the second formula is called the "Discount Factor."

9.8 TIME VALUE OF CREDITORS' INVESTED MONEY

Table 9.2 is almost identical to Table 9.1 except that it includes one more column that is titled "Present Value to the Creditors" to the left side

Table 9-02 Present Values of the Future Payments

Loan amount	$ 421,000	Interest Rate	7.50%	Term	8
Payment	$ 71,876				

Term (years)	Present Value to the Creditors	Payment (Debt Service)	Rate	Interest $	Principal	Balance
0						$ 421,000
1	$ 66,861	$ 71,876	7.50%	$ 31,575	$ 40,301	$ 380,699
2	$ 62,197	$ 71,876	7.50%	$ 28,552	$ 43,324	$ 337,375
3	$ 57,857	$ 71,876	7.50%	$ 25,303	$ 46,573	$ 290,802
4	$ 53,821	$ 71,876	7.50%	$ 21,810	$ 50,066	$ 240,736
5	$ 50,066	$ 71,876	7.50%	$ 18,055	$ 53,821	$ 186,916
6	$ 46,573	$ 71,876	7.50%	$ 14,019	$ 57,857	$ 129,058
7	$ 43,324	$ 71,876	7.50%	$ 9,679	$ 62,197	$ 66,861
8	$ 40,301	$ 71,876	7.50%	$ 5,015	$ 66,861	$ –
Total	$ 421,000	$ 575,009		$ 154,009	$ 421,000	

of the "Payment (Debt-Service)" column. The amounts shown in the column are discounted amounts that represent the PVs of the future payment creditors will receive every year.

At the end of the first year, the payment to be made to creditors will be $71,876 (shown in the "Payment" column). However, the PV of the amount at the time of investing, discounted by the risk level of 7.5% (which is the interest rate) using the formula as in "$71{,}876 \times (1\ (1 + 0.075)^1)$", is only $66,861. This can be also interpreted differently as "if I invest $66,861 now and make it grow at the rate of 7.5% for one year, it becomes $71,876." The interest rate of 7.5% represents the risk taken by the creditors that should be compensated. The amounts at two different times are different, but the value remains the same to investors when the risk is compensated. The second year's debt-service payment works the same way. Simply, creditors must wait for 2 years to have this amount. When it is discounted at the same rate (7.5%), it is equivalent only to

$62,197. If all present values are added together, the total becomes the same amount of the current debt of $421,000. For this amount, R&B Grill makes the total payments of $575,009. This means that, to the creditors, the invested amount of $421,000 is considered the same as $575,009 in value in 8 years when 7.5% risk is taken annually.

9.9 CONCLUSION: TIME VALUE OF EQUITY INVESTORS' MONEY

The interpretation provided above is very important to understand the role of a business when it finances necessary funds from investors. It takes creditors only a fixed length of time to recover their investments because long-term debts are issued with a fixed payment schedule as presented in Tables 9.1 and 9.2. The borrower must comply with the payment schedule. However, equity investors do not have payment schedules for their return. As their return is the Net Profit of the business, equity investors must wait for the net profit to be announced by the business they own. If no net profits are earned, equity investors receive no returns. Furthermore, net profits from a business cannot grow unlimitedly. These are a few reasons that equity investors have to take higher risk when compared with creditors.

Using the CAPM, equity investors estimate the reasonable rate of return on their investments and demand it to the firm they invest in. Just like creditors, equity investors also discount their expected returns from the business at the rate they have estimated.

Up to this point, discussions have been about the risk and return from the investors' perspectives. However, it is the business that must provide returns to its investors by generating enough profits. For this reason, it is important to estimate the reasonable rate of return that a business must achieve to fulfill this task of satisfying the two different demands. The next chapter finalizes this approach with a scenario of R&B Grill.

CHAPTER 10

VALUATION OF R&B GRILL

10.1 INTRODUCTION

This chapter finalizes every topic discussed so far by presenting the Present Value (PV) of a business through the process of valuation. The ultimate goal of creating and running a business is to increase the value of its owners' wealth in the long run. This can be achieved only by increasing the market value of the business. The market value of a firm is different from its book value that is presented as total assets in the Balance Sheet. The market value of a firm is the perception of investors in the market regarding the company's future growth potential. Valuation of a firm is the process estimating the amount of cash it can generate for its owners in the long run. For this reason, valuation has to start with forecasting the results of future operations that will generate cash for equity investors. This chapter will introduce an example of this process by applying all information developed up to this point and determine the true market value of R&B Grill.

10.2 REVIEW OF PREVIOUS CHAPTERS

Everything this book has introduced is divided into three parts. The first part that includes first three chapters introduced necessary accounting techniques to guide readers to understand how accounting information is generated and how it is presented in the Income Statement and in the Balance Sheet. It also introduced how these two financial statements are interrelated representing different dimensions of a business.

The next part introduced necessary techniques that analyze business performance from different perspectives. Examples were provided to guide readers to develop comprehensive views about the performance of a business by comparing different sets of accounting information with operational data. Techniques of short-term budgeting were also presented in this section.

The last part introduced the concept of cash flows as the ultimate measurement unit to value a business. Along with cash flows, technical details of long-term forecasting were introduced using the example of R&B Grill. Finally, Chapter 9 introduced the concept of risk that is an inherent condition of investing activities followed by the technical concept of the Time Value of Money (TVM) that uses the elements of risk – expectation and time – as the main criteria to determine the appropriate rate of return on investments.

This chapter presents the final framework of valuation by summarizing all the previously introduced concepts and techniques and determines whether R&B Grill, as a restaurant company, is worth for investing. In other words, this chapter will determine whether the business of R&B Grill is expected to "Add Values" to the equity holders' investments.

10.3 THE CONSOLIDATED RISK OF A BUSINESS IMPOSED BY FINANCIAL SUPPLIERS

Chapter 9 pointed out that a business must face a variety of uncertainties while it conducts its activities. Despite uncertainties, a firm must prepare necessary resources (assets) through financing from financial suppliers – creditors and equity investors (owners) – who demand specific returns on their investments and the firm must provide it. These demands of returns impose financial risk to the firm that uses the invested funds.

First, the firm must pay interest to the creditors for using the borrowed funds beyond the principal. Interests, in this context, are called the Cost of Debt Capital. On the other hand, owners' demand of return called the Cost of Equity Capital. This is illustrated in Figure 10.1: Value-adding Through Service Exchange.

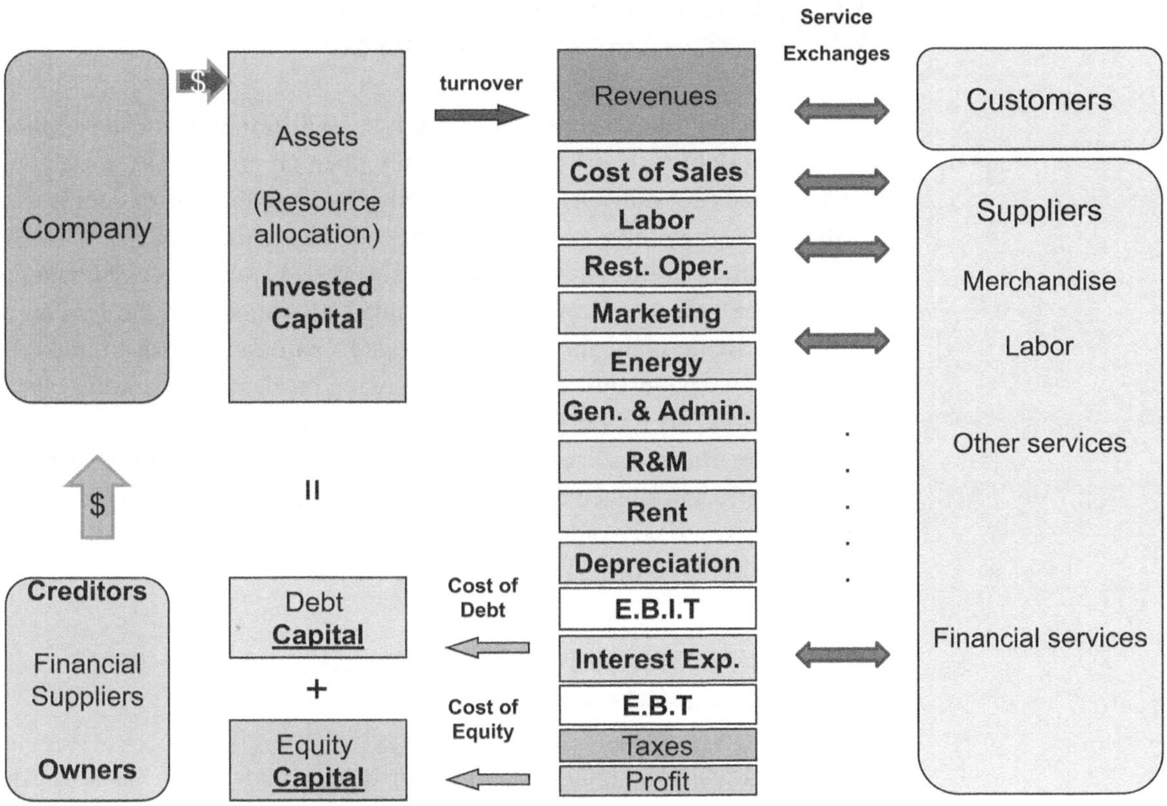

Figure 10.1 Value-adding through service exchange.

In Figure 10.1, the first column on the left shows the relationship between a company and its financial suppliers. The next column represents the accounting relationship between them as in the equation of $A = L + E$. When financial suppliers invest their funds to a company, the company recognizes them as its debt (liabilities) and equity. Both are separate capitals invested by two different groups of financial suppliers with different rates of return required. Once a firm finances its necessary funds, they become its own "Invested Capital." As the invested capital belongs to the company, it becomes its "Assets." A company must utilize its "Invested Capital" effectively and efficiently to generate enough return on its own capital use so that it can be distributed to its financial suppliers. Differently put, a firm's own ROI must be enough to satisfy the demanded ROI's of individual financial suppliers.

10.4 CAPITAL STRUCTURE AND THE WEIGHTED AVERAGE COST OF CAPITAL (WACC)

Different rates of return required by two different financial suppliers – cost of debt and cost of equity – may open an opportunity for a company to control down its financial burden because the cost of debt is usually lower than the cost of equity due to the magnitude of risk taken by the owners. For this reason, borrowing seems a better choice of financing in the first glance. However, debt capital also increases the burden to the borrowing company to prepare a large amount of cash to make its debt service payments. Despite the growing burden of generating higher amount of cash for debt payment, the advantage of debt financing still should be considered seriously. Related to this, the tax-effect of interest was briefly introduced in the previous chapter. The next section introduces a simplified example that shows more details of the benefit of borrowing (debt financing) ultimately for owners.

10.5 ADVANTAGES OF DEBT FINANCING TO THE EQUITY INVESTORS

Let's create an imaginary firm that is launching a new business. The new business requires the capital investment of $100,000. The firm has two different financing options – Option A and Option B. Option A is full equity financing with no debt and Option B is financing 40% through long-term debt at 10% interest rate, and the rest will be financed through equity.

When the new business is fully operated, its asset turnover is expected to be twice to generate annual revenues of $200,000. Its total expenses without depreciation are expected to be $160,000. The annual depreciation expense will be $10,000. The Income Tax rate is 40%.

Table 10.1 presents the pro forma Income Statement of the two scenarios accommodating all information introduced above.

As presented in Table 10.1, Scenario A with no debt provides better Net Profit and its return on sales (ROS) and return on asset (ROA) are better than those of Scenario B with debt. However, as for the ultimate goal of business that is to provide maximum return to the owners, Scenario B with debt yields better return on equity (ROE). At the same time, Scenario B spends less on the Income Tax by $1,600. This saving

Table 10-01 Advantage of Long-term Debt in the Entire Capital

Scenario		No Debt (A)	With Debt (B)
Average Total Assets	A	100,000	100,000
Average Total Debt	L	–	40,000
Average Total Equity	E	100,000	60,000
Pro forma Income Statement			
Revenues		200,000	200,000
All other Expenses		160,000	160,000
Depreciation Exp		10,000	10,000
EBIT		30,000	30,000
Interest Expenses (10%)		–	4,000
EBT (Taxable Income)		30,000	26,000
Tax (40%)		12,000	10,400
Net Profit (Return)		18,000	15,600
R.O.S (try yourself)		9.00%	7.80%
R.O.A (try yourself)		18.00%	15.60%
R.O.E (try yourself)		18.00%	26.00%

of \$1,600 on the Income Tax is exactly 40%, which is the tax rate, of the amount of Interest Expenses. As 40% of the Interest Expenses have been saved, the real interest rate has also been reduced by 40% from the original 10%. When this reduction of 40% is applied to the original 10%, the after-tax interest rate becomes only 6% (10% × (1-0.4). This will be explained again in the following section of After-Tax Interest Rate.

These are the benefits of financing through long-term debt for the owners of a company. However, this decision must be made with caution considering the payment of the principal included in the annual debt service.

10.5.1 After-Tax Interest Rate (True Cost of Debt Capital)

Thanks to the tax-deductible nature of interest, a business saves on its income tax, as illustrated in the previous section. This tax-saving effect can be applied into the interest rate and lowers it. In Scenario B of the previous example, the company's interest is 10% and pays $4,000 for its Interest expense, but saves $1,600 in its Income Tax. This saving of $1,600 is 40% of its interest amount ($4,000). With this saving included, the actual spending on interest becomes $2,400. Comparing the actual spending (after saving on tax) is 6% of the long-term debt ($40,000). This result can be formulated as follows:

$$\text{After-tax interest rate} = \text{Interest Rate} \times (1 - \text{tax rate})$$

$$\text{After-tax Interest rate} = 10\% \times (1- 0.4) \quad = 6.0\%$$

Let's apply this to R&B Grill. Its total amount of long-term debt is $240,000 at the end of the first year (Y-1). With the interest rate of 7.5%, the interest amount becomes $18,000. However, this spending reduces its taxable income 35% of the interest amount ($6,300). The after-tax interest rate of R&B Grill reduces to 4.875% from 7.5% resulting in reduced financial burden. This logic is presented in the following equation:

$$\text{After-tax interest rate} = \text{Interest Rate} \times (1 - \text{tax rate})$$

$$\text{R\&B Grill's After-tax Interest rate} = 7.5\% \times (1 - 0.35) = 4.875\%$$

The tax-deductible nature of interest expenses is very important in financing decisions. R&B Grill has reduced its financial burden of 7.5% on its long-term debt to below 5%. At the same time, by financing through long-term debt, it will provide better return to its owners as shown in Table 10.1.

10.6 COST OF EQUITY CAPITAL AND THE WEIGHTED AVERAGE COST OF CAPITAL (WACC)

The rate of return demanded by the owners of R&B Grill was determined as 5.3% in Chapter 9 applying the concept and technique of the CAPM. R&B Grill must generate enough profits to satisfy the two financial supplier groups. This section introduces final technique that consolidates two different costs into one overall cost of capital of the company.

It must be remembered that R&B Grill, as a business, uses its entire capital financed from two sources – long-term debt and equity. At the end of the first year (Y-1), the company is carrying $240,000 of long-term debt at the interest rate of 7.5% and $714,705 in its equity. Combining these two capitals, the total capital owned by R&B Grill is $954,705. This is the entire Invested Capital used by R&B Grill for its operations. The amount of total assets in the Balance Sheet is higher because it includes current assets. Current assets and current liabilities are not included in estimating the cost of capital because they are considered operating funds for short-term use rather than invested capital for long-term business.

R&B Grill's invested capital is made of about 25% long-term debt and about 75% equity. This is the capital structure of this company. It must generate 4.88% net return (rounded after-tax interest rate) for creditors and 5.3% of return for its owners as determined in Chapter 9 using the total amount of $904,105 in its business. The converged rate of return for the company can be obtained as follows:

According to the result presented in Table 10.2, R&B Grill needs to generate return of $49,645 that is 5.2% of its entire invested capital ($954,705) to satisfy the creditors and the owners by providing what they demand on their investments. The consolidated cost of using the entire capital, which is mixed with debt and equity, is 5.2%. This is called the Weighted Average Cost of Capital (WACC). The nature of this cost is almost the same as the concept of the ROA. The only difference is that the ROA uses the entire assets, while the WACC uses the invested capital. Because the entire assets are larger than the Invested Capital, if the ROA of a business is at least the same as its WACC, the business can be considered strong enough in pursuing its goal of adding value for its owners. It is also known as the company's minimum hurdle rate to satisfy its financial suppliers.

Table 10-02 Weighted Average Cost of Capital of R&B Grill – Low

Capital	Amount	Weight	Cost %	Weighted Cost	
Debt Capital	$ 240,000	25.1%	4.88%	1.2%	*
Equity Capital	$ 714,705	74.9%	5.30%	4.0%	**
Total Capital	$ 954,705			5.2%	***
Necessary amount of return on the total invested capital				$ 49,591	

* Weight (25.1%) x Cost % (4.88%) = 1.2% (Weighted Cost of Debt)

** Weight (74.9%) x Cost % (5.30%) = 4.0% (Weighted Cost of Equity)

*** Total of the two (1.2% + 4.0%) = 5.2% (Weighted Average Cost of Capital)

This result is obtained using the cost of equity capital estimated in Chapter 9. In this chapter, however, the cost of equity capital will be adjusted higher to accommodate the high risk taken by the owners for their investing into a small business. This approach should be considered more practical. The following section provides much more conservative valuation with a long-term projection.

10.7 LONG-TERM PROJECTION OF R&B GRILL FOR VALUATION WITH HIGHER COST OF EQUITY CAPITAL

This section provides the final valuation with a long-term projection of R&B Grill as presented in Table 10.3.

Table 10.3 is almost the same as Table 8.6. It shows the details of the projected operating results of the next 7 years when the business is expected to mature and stop growing. Underneath the pro forma, Income Statements are a few sets of supporting information that are used in valuation process.

Section I shows the Balance Sheet information in the beginning of Year 2 (Y-2). Assets and liabilities are larger than the total amount of Invested Capital and that of the long-term debt as explained earlier. It is assumed that R&B Grill will maintain this financial position in the future.

Section II presents the summary of ratio analysis of the projected operating results. The ratio analysis has adopted the total amounts of assets, liabilities, and equity from Section I to present more conservative estimates. All ratio analyses show positive results in the future. When compared with well-known chain restaurant companies that have similar concepts, such as Bloomin' Brands, Brinker International, and Darden, the results of R&B Grill are a little lower than those of its nation-wide competitors. So, it can be concluded that this projection is conservative and reliable.

Section III is the projection of cash flows in a simplified format. The Cash Flows from Operations are the net profit of each year added with the amount of depreciation expenses. The Cash Flows from Investing are estimated amounts that the company will have to spend to sustain its business. As time goes by, R&B Grill will have to spend more to maintain its assets in good shape. These amounts also include necessary

Table 10-03 Long-term Projection and Analysis

The Pro Forma Income Statements of the next 7 years The R&B Caterer, Inc.	FC	VC/ cust	Y-2	V.A.%	Y-3	Y-4	Y-5	Y-6	Y-7	V.A.%
	Customer #		58,604		60,362	62,173	64,038	65,959	67,600	
	Avg. Check*		$26.35		$26.85	$27.35	$27.90	$28.45	$29.00	
	Increase % (exp. Inflation)				1.9%	1.9%	2.0%	2.0%	1.9%	
Revenues			1,544,215	100.0%	1,620,723	1,700,431	1,786,665	1,876,543	1,960,400	100.0%
Cost of sales	Cost %	33.5%	517,312	33.5%	542,942	569,644	598,533	628,642	656,734	33.5%
Gross Profit			1,026,903	66.5%	1,077,781	1,130,787	1,188,132	1,247,901	1,303,666	66.5%
Operating Expenses	FC (annual)	VC/cust								
Labor expenses	per labor schedule		507,395	32.9%	517,543	527,894	538,452	549,221	560,827	28.6%
Restaurant operating expense		$ 1.25	73,255	4.7%	75,453	77,716	80,048	82,449	84,500	4.3%
Marketing expenses	$ 20,000		20,000	1.3%	20,000	20,000	20,000	20,000	20,000	1.0%
Energy (Utility) expenses	$ 11,652	$ 0.41	35,680	2.3%	36,400	37,143	37,908	38,695	39,368	2.0%
Administrative expenses	$ 58,980	$ 0.72	101,175	6.6%	102,441	103,745	105,087	106,471	107,652	5.5%
Repair & maintenance expenses		$ 0.30	17,581	1.1%	18,109	18,652	19,211	19,788	20,280	1.0%
Rent expenses	$180,000	$ –	180,000	11.7%	180,000	180,000	180,000	180,000	180,000	9.2%
Depreciation – FF&E	$ 24,000	$ –	24,000	1.6%	24,000	24,000	24,000	24,000	24,000	1.2%
Depreciation – Leasehold improve.	$ 24,000	$ –	24,000	1.6%	24,000	24,000	24,000	24,000	24,000	1.2%

Table 10-03 (Continued)

Depreciation – Smallware	$ 24,000	$ –	24,000	1.6%	24,000	24,000	24,000	24,000	24,000	1.2%
Total Operating expenses			1,007,086	65.2%	1,021,945	1,037,149	1,052,706	1,068,624	1,084,627	55.3%
Operating profits			19,818	1.3%	55,835	93,637	135,426	179,277	219,039	11.2%
Interest expense (income)			18,000	1.2%	18,000	18,000	18,000	18,000	18,000	0.9%
Income before income taxes			1,818	0.1%	37,835	75,637	117,426	161,277	201,039	10.3%
Income taxes (35%)			636		13,242	26,473	41,099	56,447	70,364	
Net Income (loss)			1,181	0.1%	24,593	49,164	76,327	104,830	130,675	6.7%

Section I

Balance Sheet records in the beginning of Y-2: R&B Grill's financial positions will be maintained at the current values.

Average Assets	1,135,705
Average Liabilities	421,000
Average Equity	714,705

Section II

Ratio Summary – projection	Y-2	Y-3	Y-4	Y-5	Y-6	Y-7
Assset Turnover	1.36	1.43	1.50	1.57	1.65	1.73
Return on Sales	0.08%	1.52%	2.89%	4.27%	5.59%	6.67%

	Return on Assets	0.10%	2.17%	4.33%	6.72%	9.23%	11.51%
	Return on Equity	0.17%	3.44%	6.88%	10.68%	14.67%	18.28%
Section III	Cash Flows Projection						
	Cash Flows from Operations	73,181	96,593	121,164	148,327	176,830	202,675
	Cash Flows from Investing	20,000	30,000	40,000	50,000	60,000	80,000
	Free Cash Flows	53,181	66,593	81,164	98,327	116,830	122,675
Section IV	Annual Debt Service Information:						
	Long-term Debt		$240,000				
	Interest Rate		7.50%				
	Terms to pay off the entire debt		8				
	Annual Debt Service amount		40,974				

* Annual Free Cash Flows are sufficient to cover the Debt Service.

spending for more inventory purchases and other expenses to support increasing volume of its sales. The Free Cash Flows in the last line is the difference between the two above (Cash Flows from Operations – Cash Flows from Investing). By the definition, this is the amount of remaining cash in the company that can be freely used by the management. However, the payment of the long-term debt and interest that is explained in the following section must be made out of this cash.

The next section (IV) shows simplified information of R&B Grill's annual debt-service amount. The company has long-term debt of $240,000 in the beginning of Y-2. The interest rate was set at 7.5% and the term is 8 years. To pay off this debt, the company must make the annual payment of $40,974 for its principal and interest together. According to this projection, even the second year's (Y-2) Free Cash Flows are enough to pay for the annual debt service. Even if the actual performance should fall short of this projection, R&B Grill will be able to make the payment by tapping into its a large amount of cash reserved for emergency situations as explained in an earlier chapter. Keeping extra amount of cash on hand for contingency is always recommended for a business to maintain its financial flexibility. In conclusion, R&B Grill looks strong in its operations for its financial dues according to the current analysis and projection.

10.8 FINAL VALUATION OF R&B GRILL WITH A REVISED WACC: IS IT WORTH INVESTING?

This is the final part of valuation that determines whether the business of R&B Grill is a good opportunity for its owners to invest. In the beginning, three owners have invested $650,000 altogether with additional long-term debt of $400,000. Since then, throughout the entire book, a variety of scenarios and relevant transactions has been introduced and consolidated into the Income Statements and the Balance Sheet. Now that the future performance is projected into pro forma Income Statements with other necessary information, such as cash flows and the cost of capital, it is time to determine whether this business is truly a good opportunity for its investors.

To make this valuation more reliable, the cost of equity capital will be revised higher as suggested earlier. The higher demand of return is reflected in the perceived risk level (beta) of the business, and in the overall rate of return from the investments into the business sector instead of the governmental bonds (Risk Premium). We will accommodate the higher demand and increase the beta of R&B Grill to 1.8. The Risk Premium (or Normal Market Premium) will be adjusted from to 5%. The results of these adjustments are presented in Table 10.4, followed by explanation.

The first section (I) of Table 10.4 shows necessary information to calculate the rate of return demanded by the owners of R&B Grill and the WACOC. The first line shows the interest rate of 7.5%. This is reduced to 4.88% with the tax effect accommodated. The next line shows two different pieces of information of the risk-free rate, which is the yield rate of 10-year Treasury Bonds at the time of writing this script, followed by the Risk Premium. The Risk Premium has been adjusted to 5% considering the higher risk involved in a small private restaurant business. The next part of Section (I) lists the beta values of major nationwide chain restaurant companies that are considered direct competitors to R&B Grill. Compared with the beta values of competing companies, the beta of R&B Grill, estimated at 1.8, presents much higher risk involved. This means that R&B Grill is perceived 1.8 times as risky for investment as the entire market. Using the CAPM formula, it turns out that R&B Grill must provide 10.7% of return to its owners for their investment. In other words, the owners of R&B Grill should have 10.7% return on their investment.

Section II shows the details of R&B Grill's WACC calculation. Mixing the two capitals into its entire invested capital, the revised WACOC is 9.2%. This is almost twice as high as the previous WACC of 5.3%. This is the rate of return R&B Grill must generate using its entire capital of $954,705.

Section III presents projected Free Cash Flows as presented earlier in Table 10.3. If R&B Grill performs as expected, its Free Cash Flows will be as shown on the third line of Section III "Valuation Table." As mentioned, these are the amounts of cash the company uses to pay for its long-term debt.

As the table presents, the amount of Free Cash Flows grows each year and by Y-7, it is expected to stop growing. It is natural that any business stops growing after a reasonable amount of time. From there on, the

Table 10-04 Final Valuation of R&B Grill

I WACC Calculation

Interest Rate	7.50%	After-tax Interest Rate:	4.88%	
10-year T-bill rate	1.70%	Risk Premium	5.00%	Increased

Beta of similar companies for reference:

Bloomin' Brands	0.80	
Brinker International	0.32	
Darden	0.52	
R&B Grill's self-estimated risk (beta):	1.80	Increased
Capital Asset Pricing Model (Rf + Beta x RP)	10.7%	

II Weighted Average Cost of Capital Table

Capital	Amount	Weight	Cost %	W. Cost
Debt	240,000	25.1%	4.88%	1.2%
Equity	714,705	74.9%	10.70%	8.0%
Total	954,705		WACC	9.2%

III Valuation Table:

Year	Y-2	Y-3	Y-4	Y-5	Y-6	Y-7
# of years to wait	1	2	3	4	5	6
Free Cash Flows	$53,181	$66,593	$81,164	$98,327	$ 116,830	$ 122,675
		Terminal Value at the end of Y-7 when growth stops				1,333,424
WACC	9.2%	9.2%	9.2%	9.2%	9.2%	9.2%
PV of FCF	$48,701	$55,845	$62,330	$69,148	$75,239	$ 786,380

True Marketl Value of the Company's Assets in the beginning of Y-2		$1,097,642
Less the PV of the liabilities in the beginning of Y-2		$ 400,000
The Present Value of the company's Equity in the beginning of Y-2		$ 697,642
Less the Total amount of Equity in the beginning of the Y-2		$ 650,000
Net Present Value (NPV) of R&B Grill at the beginning of Y-2	Positive	$ 47,642

business will experience stable performance for a while before it eventually declines. In valuation, it is anticipated that the same amount of Free Cash Flows will be continuously generated from that point without declining. This anticipation is based on an assumption that new concepts will be designed and implemented to meet with the emerging trends of the business environments and new growths will take over the decline. This "life-cycle" concept is presented in Figure 10.2.

It is time to measure the true value of R&B Grill by consolidating the projected future Free Cash Flows into the PV in the beginning of Y-2. At the time, the company had $1,128,300 invested into its total assets. By using this amount, R&B Grill must perform as projected in the pro forma Income Statement and earn the Free Cash Flows as presented in Table 10.4.

The last Section (III) "Valuation Table" in Table 10.4 shows the annual amount of Free Cash Flows until the company matures. Our projection estimates that R&B Grill's current concept will mature in the seventh year (Y-7) and from that year its Free Cash Flows will remain the same amount forever. The line above the FCF shows the number of years

- A business – products & services – goes through its own life cycle of:
  - Starting *(a large amount of investments is needed.)*
  - Rapid growth *(business picks up, cash flows are increasing...)*
  - Slow growth into mature stage *(growth slows down)*
  - Decline
  - *In valuation, when the business matures, it is assumed to last with no changes.*

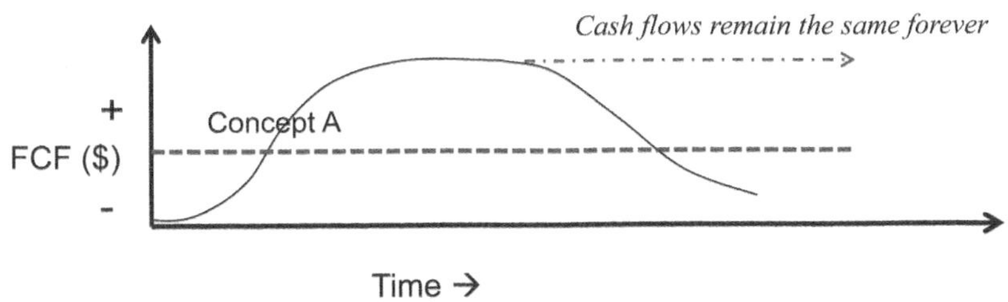

Figure 10.2 Valuation with the life-cycle theory.

for investors to wait to have each year's projected Free Cash Flows. For example, in the beginning of Y-2, investors must wait 1 year to have the Free Cash Flows (FCF) of the year. They have to wait for 2 years in the beginning of Y-2 to have the FCF of Y-3, and so forth. The line below the FCF says "Terminal Value at the end of Y-7 when growth stops." This concept needs further explanation as provided in the next section.

10.8.1 Terminal Value

As explained, this (Y-7) is the year when the company's business concept matures, and R&B Grill stops growing. Its Free Cash Flows will remain at the same amount ($122,675) forever. When the same amount is generated forever, it is called "perpetuity," and its PV is calculated as follows:

Present Value (PV) of Perpetuity = One amount/Expected Rate of Return

Applying the formula above, the PV of R&B Grill's future FCF from Y-7 can be obtained as follows:

Present Value (PV) of Perpetuity = One amount/Expected Rate of Return

PV of FCF of Y7 = One amount $(122,675)/WACC (9.2%)
= $1,333,424

(* *The result of calculation may slightly differ depending on the calculator setup. Readers are recommended to use their personal judgment.*)

The result of $1,333,424 is the PV in the beginning of Y-7 of all future perpetuity of $122,675 from that year. It means that R&B Grill will carry its market value at $1,333,424 in the beginning of Y-7 according to our projection. It must be remembered that this amount ($1,333,424) is still the FV that will happen in 6 years from the beginning of Y-2.

The PV of the perpetuity from Y-7 ($1,333,424) must be discounted one more time to the PV in the beginning of Y-2 (which is $786,380), and all other future FCF's must be discounted to the value of Y-2 to estimate the market value of R&B Grill. The next line in Section III "PV of

FCF" of Table 10.4 shows the result of this process. When all these PV amounts (in the beginning of Y-2) are combined, the total value of the company is $1,097,642.

The market value of R&B Grill estimated at $1,097,642 represents the capability of its total assets utilized by the management to generate cash for the entire life of the business. From this, the amount of total liabilities, $400,000 in the beginning of Y-2, must be taken out to measure the market value of Owners' Equity. The result is $697,642. This is the market value of Owners' Equity of R&B Grill in the beginning of Y-2.

10.9 CONCLUSION

The market value of the equity of R&B Grill ($697,642) introduced in the previous section needs more explanation. It is the difference between the market value of the company's assets and the total amount of its liabilities. As both – the assets and the liabilities of the company – are in present values at the time of valuation, the difference between the two (owners' equity) must be in PV. Therefore, it represents the estimated market value of the equity. If R&B Grill achieves the goal of generating Free Cash Flows as projected according to the scenarios presented in this book, the original amount ($650,000) of owners' investments in the beginning of the Y-2 is worth $697,642.

It is also important to analyze the final amount of $697,642. It is the discounted amount of all future cash flows accommodating the risk taken by investors. For this reason, this amount can be compared fairly with that of the invested amount of $650,000. The concept of R&B Grill and its operating capability within its business environment will add the net value of $47,642 to the investment of $650,000. In conclusion, the business of R&B Grill is worth investing because it adds value to the original investments even after compensating the high level of risk accommodated into the scenario.

10.10 NET PRESENT VALUE (NPV)

The technique and the concept used in this framework are called the Net Present Value (NPV) approach. As the title suggests, this approach estimates all possible future earnings (in Cash Flows from Operating

Activities) and discounts them to the PVs. Simultaneously, it also estimates all the necessary spending of the future and discounts them to the PV. Then, from the PV of future earnings, it takes out the PV of all future spending. The difference is the NPV of a business. If the NPV is positive, the business adds value to the equity holders' investments. It is recommended to make investments into the business. Otherwise, it fails to add value; therefore, it is not recommended to invest into the business.

As an additional introduction, in the case of a public company, the most reasonable stock price can be determined by dividing the PV of the owners' equity by the number of outstanding common stocks in the market. If the estimated price is higher than the market price, the company is undervalued; therefore, it is recommended to buy its stocks. Otherwise, it is recommended to sell them.

APPENDIX

I. ACCOUNTING PRINCIPLES (GENERALLY ACCEPTED ACCOUNTING PRINCIPLE: GAAP)

The discipline of Financial Accounting has advanced to the highly structured mechanism to provide accurate and reliable information of business for a variety of user groups. In the root of the accounting method is a set of generally accepted principles used as a uniform basis for preparing financial statements. These principles, called Generally Accepted Accounting Principle (GAAP), also have evolved through the work of such major accounting bodies as the American Institute of Certified Public Accountants (AICPA) and the Financial Accounting Standard Board (FASB). Although hospitality businesses are not legally required to follow them because these principles are not laws, there is a strong presumption of misrepresentation of the financial statements if a firm has not followed the principles. In case of a public company, any significant departure from GAAP must be specified in the CPA's opinion letter that is included in the annual report so that public can be aware of them.

In Chapter One, three of GAAP – the principles of Business Entity, Cost, and Objective Evidence – have been introduced with relevant transaction examples. Among many principles, most Financial Accounting textbooks of the Hospitality Management introduce the following ones:

- Business Entity Principle (or Economic Entity Principle)
- Cost Principle (or Historical Cost Principle)
- Objective Evidence Principle
- Unit of Measurement Principle
- Continuity Principle (Going Concern Principle or Concept)
- Conservatism Principle
- Full Disclosure Principle
- Consistency Principle
- Matching Principle
- Materiality Principle
- Revenue Recognition Principle

The following are the explanations of each principle with more conceptual background.

- **Business Entity Principle (Economic Entity Principle)**

As explained in Chapter One, this principle states that a business is considered an independent entity (an economic unit), for accounting purposes, that maintains its own set of accounts; and these accounts are separate from other financial interests of the owners. Under this principle, if the owner of a business takes the property of the firm for personal use, it should be properly charged to a separate account of the owner. This separate recording not only provides accurate and reliable information for managing the business, it is also necessary for properly filing tax returns of the business. Another example includes the debts of a business. They must be reported separately from the debts of the owner. This means that only those debts that are directly related to the business should be included in the financial statements of the business.

- **Cost Principle**

According to this principle, only the transaction price (actually paid amount paid for the purchase) becomes the accounting value for the product or service purchased that becomes an asset item for the business. At the time of purchase, the price suggested by the manufacturer (MSRP) or even the market value is irrelevant. The seller of the product, on the other side of the transaction, also records the actual transaction amount for its accounting record. Only by doing so, the amount becomes objective and fair measure of the value of the item purchased. This way, cost can be defined as the price paid to acquire an asset; and it is the proper basis for accounting for the resources owned by the hospitality organization. The acquisition cost of an asset remains in the accounting records until the asset is sold, expires, or is consumed by the business.

This concept is very closely related to the continuity principle (going-concern concept), which is explained further later in this section. Furniture, Fixture, and Equipment (FF&E) of a hospitality business is normally carried at cost less depreciation. This treatment of property and equipment is based on the continuity (or going-concern) principle.

Additionally, when the value of an asset declines to lower than the cost recorded on the books, the declined value must be presented in the accounting book (the Balance Sheet) according to the conservatism principle. Thus, the conservatism principle overrides the cost principle. However, many hospitality businesses do not reduce their FF&E to market value as long as management plans to retain them for their useful lives.

■ Unit of Measurement Principle (Money Measurement Concept)

All accounting information must be expressed in monetary terms. In the United States, the unit of measurement is the U.S. dollar. Only with such standard, accounting reports would be meaningful. In other words, such information as two buildings and/or 500 acres of land, with no monetary value is not considered meaningful in accounting. Only the monetary value is assumed to represent a stable measurement of value that makes historical transactions from past period and the current period comparable on the same statement. However, unit of measurement principle does not explain the decreasing purchasing power (time value) of money over time. This part will be further explained in the last part (Part III).

■ Objective Evidence Principle

All accounting information, such as transactions and the resulting accounting records, must be based on objective evidence as possible. This evidence is usually an invoice or a canceled check that verifies the record of transaction. Sometimes, however, assumed estimates may be allowed in the absence of hard evidence. In such occasions, due to the judgmental value that are often estimated for assets on financial statements, there must be a systematized uniform method so that others repeating the same estimate at different times will obtain the same results.

When estimates are used to measure the value of some transaction, conservatism principle mentioned earlier must be used to support the objective evidence. Schmidgall and Damitio introduce an example as follows in their book (Hospitality Financial Accounting, 4th ed., published by the American Hotel & Lodging Educational Foundation, 2015).

"If the owner of a restaurant contributes equipment, purchased several years ago for personal use, to a restaurant corporation in exchange for 100 shares of stock. Further assume that there is no known market value for the restaurant corporation's stock. The owner may believe the equipment is worth $1,000, while the original catalog shows the cost several years ago of $1,400, and an appraiser places the value of the equipment at $850. In this example, the most objective estimate of its value today would be the appraiser's estimate of $850."

■ Continuity Principle (Going Concern Concept)

In preparing the accounting records and reports, it must be assumed that the business will continue indefinitely with no planned liquidation in foreseeable future, which, in conclusion, expects that the business is a going concern. This principle is based on the concept that the real value a business is its capability of earning profits long into the future, assuming that the resources are used in the regular operations of the business.

According to this concept, since the business is not to be sold or liquidated, there is no need to record current replacement values (market values) or liquidating values. Hence, the current resale value of these resources is irrelevant in the measurement process. If there is a reliable reason, however, the business may be unable to continue its operations in the near future, allowance for this future event should be reflected in financial statements. Put differently, when there is a good reason to believe that the continuity principle is no longer applicable to a particular firm, then management is required to measure the resources owned by the business at their liquidating value. Because presenting the liquidation value of assets requires a firm to in partial or total liquidation, using liquidation values on financial statements is seldom justified.

■ Conservatism Principle

In a technical sense, conservatism means using the measurement method that yields the least favorable immediate results. This approach reflects the working rule, "Do not create a false hope," or "Anticipate no gains, but provide for all possible losses." Thus, this principle calls for minimizing gains and maximizing losses by recognizing expenses as soon as possible, but delaying the recognition of revenues until they are ensured. By doing so, the result is to be conservative (low) in recognizing net income in the current year.

When this principle is applied to a lawsuit, a business should recognize the expenses immediately if it is a defendant and it is expected to lose the lawsuit. On the other hand, if a business is a plaintiff in a lawsuit and is expected to win the case, the estimated amount of settlement is not recorded until a judgment is rendered.

Conservatism is apparent in the valuation of inventory at the lower of cost or current market. The use of a liability account of "Unearned revenues" or "Advanced deposits" for future banquets until the service is rendered is another example.

However, it must be emphasized the conservatism principle should not be used to justify an unwarranted reduction in the asset cost and earnings of a business. An example of this is the specific area of future events. If two estimates of future payment, one greater than the other, are equally likely to occur, conservatism requires the use the lesser estimate. If, however, the greater future payment estimate is more likely to happen, it will be more objective to use the greater estimate rather than the lesser.

In conclusion, the conservatism principle must be applied with prudent judgment in evaluating the uncertainties and risks that are inherent in business situations. A deliberate understatement of the assets and/or profits of the firm should be avoided when it goes contrary to the facts.

■ Full Disclosure Principle

The full disclosure principle requires to provide necessary explanation on all information presented on the financial statements. This can be accomplished either by reporting the explanation in the body of the financial statements or in the footnotes to the financial statements. Necessary explanation may include, but is not limited to, the accounting methods used, changes in the accounting methods, contingent liabilities, events occurring after the financial statement date, and unusual and nonrecurring items. The next section of this appendix includes a few examples of such necessary explanation of accounting methods.

■ Consistency Principle

Some accounting information can be obtained through a few different methods. Good examples include determining the value of inventory on hand, and determining the amount of depreciation expenses on fixed assets. Depending on the specific method, the value of inventory on hand will vary; and so will the amount of depreciation.

According to the consistency principle, once an accounting method for a specific account is adopted, the same method must be used continuously unless a change is warranted and disclosed. This way, a user of financial information can make reasonable comparisons between periods. Without this principle enforced, the results presented on the financial statements might be misleading. When a company is allowed to change to another method, the change must be reported following the full disclosure principle.

■ Matching Principle

The matching principle refers to present expenses to the revenues generated during the same accounting period. A very useful example is the case of purchasing long-term assets. Long-term assets are used for more than one year to generate revenues. If a firm purchases a set of equipment for $500,000 in cash and uses it over the next 10 years, the purchasing amount (cost) cannot be allocated only to the first year because the same equipment will continuously be used to generate revenues in the following years. The cost is therefore recorded a fixed asset and the amount is written off over the system's life. By doing so, the result is a partial expense of each year against the revenues generated during the same year. More explanation will be available on depreciation later in this appendix.

Another example is the occurrence of Accounts Receivable that turns out to be uncollectable. When a business realizes that a certain amount of outstanding receivable becomes uncollectable, the amount must be presented as Bad Debt to write off the Account Receivable. When this happens, by the way, the uncollectable amount of Account Receivable may have been reported as revenues during the previous accounting period. The result of applying the uncollectable amount to the expense of current accounting period will violate the matching principle. To avoid this possible violation, a solution has been designed to use a Contra-Asset account called "Allowance for Doubtful Account (ADA)." More explanation on this case will be provided later in this appendix.

■ Materiality Principle

The term "material" in according means that events or information must be significant enough to make difference to user of the financial statements. Under this principle, an item is considered material in comparison to a standard set by management. It is reported in Schmidgall's book that some accountants have tried to establish materiality standard by rule of thumb; "for example, an item must be recognized if it exceeds a certain percentage of total assets or total income." However, this approach apparently has reportedly failed to address an item's relative importance over time; and reported some immaterial items material when viewed collectively.

Common practice of applying this principle is found in the use of "Petty Cash" account in many businesses. When there is a need to purchase miscellaneous asset items that are not material enough, the purchase is considered an expense rather than putting the item as asset and start depreciating. For example, if a microwave oven ($100 in retail price) used in an office goes out of order, and has to be replaced, the office manager can use his/her petty cash to buy a new one. The purchase is recorded as a petty cash expense; and petty cash decreases. This way, the microwave oven, which is technically a long-term asset item, does not have to be recorded on the Balance Sheet and to be depreciated over time. Most businesses have their own internal rule to determine the maximum amount of purchase that can be processed this way.

- **Revenue Recognition Principle**

The revenue recognition principle should not be confused with the way revenues are recorded as introduced in the previous chapters. Revenues are recorded only when services are rendered with the delivery of products to customers. This practice is the based on the "Accrual Basis Accounting" in comparison with the "Cash Basis Accounting."

The revenue recognition principle is about how to differentiate the price of service received from the customer. Revenues are recognized only when the business firm has provided the service as its major line of operations. For example, if a motel provides laundry service within its lines of operations and provided laundry service to its guest, the fee for the laundry should be recognized as revenues. In such situation, the motel is considered the "principal" of the service. However, if the motel has to send the laundry to external business and collects the fee with service charge, the motel can only recognize the "service charge" as its net revenues. The entire amount of the fee that must be given to the external laundry business is not the motel's revenues. In the situation like this, the motel is considered an "agent." In conclusion, if the business firm is acting as the agent, it records the net revenue (service charge only); if the hotel is the principal, however, it records the gross revenue (entire amount of the fee).

As briefly explained so far, the recording procedure of Financial Accounting must comply with all relevant principles to guarantee the accuracy and reliability of the information it generates. Only when these rules are fully accommodated, user groups of accounting information – managers, owners, creditors, and the governments – are able to trust the financial statements for their decision making.

II. HANDLING VARIATIONS OF ACCOUNTING ROUTINES

This section introduces technical details of handling irregular business activities. These are not introduced in the text due to the irregular nature that does not cause material impacts on the operational performance. However, it is recommended for operators to understand the concept of these so that they can prepare their business plans by incorporating them.

A. Writing Off Uncollectable Amount of Accounts Receivable

Managing the balance of Accounts receivable (A/R) is an important task for operators to collect outstanding cash of the uncollected revenues. Regardless of the effort to collect unpaid fees, a certain amount, if not big, may become uncollectable due to a variety of reasons. Restaurateurs must be prepared to accommodate such situations. Since Account receivable is an asset account, when a certain amount becomes uncollectable the balance of the account must be reduced. This is based on the Conservative Principle of GAAP, as explained earlier.

The method of reducing the value of the A/R is conducted through presenting the uncollectable amount as an expense item (called the "Bad Debt Expense"). This works just like the method of recording Depreciation Expenses. Just as the Depreciation Expense reduces the current book value of a long-term asset, Bad Debt expense reduces (or writes off) the amount of the book value of A/R account. These expenses (Bad Debt or Depreciation) causes the reduction on the Net Income by the same amount. When the reduced amount of the Net Income is eventually accumulated to the Retained Earnings on the Equity section of the Balance Sheet, the total amount of the Liabilities and Equity becomes the same as that of the Total Assets.

This method of reducing the balance of A/R account may violate the Matching Principle because the accounting period that has created the records of revenues and the writing off the uncollectable amount of the A/R. When a business decides to process the uncollectable amount, usually more than several months would have been passed from the revenues recorded. In such a situation, the Bad Debt recorded to write off the A/R account will have to be matched to the current accounting period's revenues amount. This is apparently a violation of the Matching Principle. To eliminate this potential problems, accountants have developed another Contra Asset account called the Account for Doubtful Account (ADA). ADA is used as a buffer account to absorb possible Bad Debt in accounting practice. Almost all publicly traded restaurant companies provide the amount of their ADA in the Footnote section in their Form 10-K Report. This is also to comply with the Full Disclosure Principle.

B. Methods of Inventory Valuation

Let's assume a scenario regarding inventory control. If a restaurant starts its business in a year with no beginning inventory. During the first year, the restaurant has purchased a product (A, for example) twice. First purchase was 200 units at $10 per unit; and paid the total amount of $2,000. The second purchase was 100 units at $15 per unit; and paid $1,500. The total purchases were $3,500 altogether. The result of the physical inventory report says there are 50 units left. Which price should be used to calculate the amount of your physical inventory – $10 per unit or $15 per unit? Two different prices will provide two different amounts for your ending inventory. This will, in turn, change the Cost of Sales Amount. When the Cost of Sales changes, the Net Profit will change, too.

To ward off the dilemma like this, a few different methods have been developed. Each method will be briefly introduced later in this section. Due to the "Consistency Principle," however, once a method is chosen, the same method must be continuously used. If the firm wants to change its inventory valuation method from one to another, it has to be reported to the government and obtain permission. Besides, the firm must report all impacts created by the change of its inventory valuation method on its annual report to comply with the Full Disclosure Principle.

There are three different methods to value the amount of physical inventory: FIFO (First-in, First-out) method, LIFO (Last-in, Last-out), and finally Weighted Average method. FIFO assumes that merchandise purchased first has been sold (used) first. LIFO, on the other hand, assumes the merchandise purchase last has been sold first. Restaurateurs must be aware that these methods are used only in accounting procedure to present the most accurate, consistent, and reliable information. Operators always use FIFO method for their daily operations to maintain the freshness of their products. Weighted Average method uses the average price. Most publicly traded restaurant companies report their inventory using the FIFO method. This must be reported in the Footnote section in their Form 10-K Report.

The following is the example of each method. Two purchases have been made: 200 units for $2,000 (unit price $10) for the first purchase, and 100 units for $1,500 (unit price $15) for the second purchase. Since the ending inventory is 50 units, 250 units have been sold.

Under the ***FIFO system***, the 250 units sold must include the first purchase of 200 units at $10 per unit; and another 50 units that were purchased at the second purchase at $15 per unit. The amount of sold units (Cost of Sales) can be calculated as follows:

Beginning Inventory + All Purchases – Ending Inventory
= *Cost of Sales* (Used Inventory)

–0- + ($2,000 + $1,500) – 50 x $15 (2nd purchase) = $2,750

Under the ***LIFO system***, the 250 units sold must include the entire last purchase of 100 at $10 per unit; and another 150 units of the first purchase at $15 per unit. Thus, the 50 units left must be the first purchase at the unit cost of $10. The amount of sold units (Cost of Sales) can be calculated as follows:

Beginning Inventory + All Purchases – Ending Inventory
= *Cost of Sales* (Used Inventory)

-0- + ($2,000 + $1,500) – 50 x $10 (1st purchase) = $3,000

As shown above, two different methods have provided two different amounts of Cost of Sales (COS). They also have provided two different amounts for the Ending Inventory. These differences obviously affect the Income Statement and the Balance Sheet. Understanding these impacts is the key elements in learning these techniques. Each method has its own pros & cons. As explained earlier, businesses must use one method consistently for their inventory valuation.

C. Methods of Estimating Depreciation Expenses

Long-term assets, when purchased, do not incur expenses to the business. The obtainment of them is considered an “Investment Activity (=Capital Expenditure).” However, as they are used in the business, they start losing their value. The business estimates this “lost value” periodically, and record it as an expense (=Depreciation Expense). When this expense is recorded (on the Debit side), "Accumulated Depreciation" is credited to indicate the Long-Term Asset item has lost that much value.

Although the Straight Line method is most commonly used, there are a few other methods to calculate depreciation. Usage method (also known as Units of Production method) is another widely used technique for such equipment as vehicles, aircrafts, copy machines, and other similar items. The other technique is called the "Accelerated Method," because this method accelerates the depreciation of the item in the beginning years. Thus, larger amounts of depreciation are recorded in the earlier stage. There are two different sub-methods under this technique – Double Declining method and Sum-of-the-years' Digit method. Both, as explained, generates larger amounts of Depreciation Expense in the beginning years.

Depreciation is a very unique expense account in that it does not involve "Cash" payment at all. All other expenses are either paid in cash immediately or they become "Accrued expenses (liabilities)" for later payments. However, Depreciation Expenses always bring up "Accumulated Depreciation" on the Credit side. By directly reporting the impact of decrease of relevant Asset items on the Credit side, no Cash payments are implied at all. For this nature, Depreciation is called "Non-Cash Expense." This is a very important and also very interesting concept of conducting a business with heavy capital investments.

Although the spending for capital investments (Capital Expenditure) is not an expense, still it causes cash to flow out (either immediately or later when payments are made for loans for the purchase). All investors want to have this spending recovered as quickly as possible through operations. Think of the following situation: Invested Capital into long-term assets depreciates in time; depreciation is presented as an expense on the Income Statement, and it reduces Net Profit, However, the depreciation does not involve real cash payments. Since there are no cash payments for the depreciation expense, the amount remains in the company in cash. Thus, the more depreciation expense is reported, the more cash remains in the company after each year's operations. For this reason, the more depreciation expense is reported, the quicker the business and the investors will be able to have their invested amount recovered. The following section explains the calculation and the impacts of different methods of estimating depreciation on tax, net profit, and on cash earned through operations using a simple scenario of a vehicle purchase.

Let's assume a scenario of purchasing a vehicle at $13,000. It will be used for 5 years. At the end of the 5th year, its salvage value is estimated to be $1,000. Additionally, projected operational performance of the future 3 years are as follows:

- Revenues will be $100,000 in the first year; $120,000 and $140,000 in the 2nd and the 3rd year.
- Cost of Sales (C.O.S) will be $30,000; $40,000; and $50,000 in the next three years.
- Operating Expenses (excluding the depreciation) will be $50,000; $60,000; and $70,000 in the next 3 years.
- Tax rate is 40% of the Earnings before Tax).

1. Straight Line Method

Since the calculation method of Straight Line method is introduced in Chapter 1, this section will directly use the annual depreciation amount of $2,400. Table PA-01 shows the projected result of the operations. Even when sales & other expenses change, the annual depreciation expenses remain the same. This scenario provides the same amount of Taxable Income (EBT) for the next three years.

The amount of the non-cash expense (depreciation) must be added back to the amount of EBT to present the amount of cash earned through the operating activities of each year. The will be explained further with more details in Chapter 7. When tax is not considered, the total amount of cash earned is $20,000 per year. This amount must be compared with other scenarios that use different methods.

2. Unit of Production (or Usage) Method:

Applying the same conditions used in the Straight Line method, let's take the Unit of Production Method. This method depreciates the asset item by the mileage driven. If the vehicle depreciates $0.12 per mile

Table PA-01 The Impact of Depreciation on Tax, Profit, and Cash – Straight Line

With no tax	Year 1	Year 2	Year 3
Revenues/Sales	1,00,000	1,20,000	1,40,000
Cost of Sales	30,000	40,000	50,000
Gross Profit	70,000	80,000	90,000
Expenses...	50,000	60,000	70,000
Depreciation	2,400	2,400	2,400
EBT	17,600	17,600	17,600

Cash Amount Earned through Operation – No Tax			
EBT (Taxable Income)	17,600	17,600	17,600
Add Depreciation exp.	2,400	2,400	2,400
Cash from Operations	20,000	20,000	20,000

With tax	Year 1	Year 2	Year 3
Revenues/Sales	1,00,000	1,20,000	1,40,000
Cost of Sales	30,000	40,000	50,000
Gross Profit	70,000	80,000	90,000
Expenses...	50,000	60,000	70,000
Depreciation	2,400	2,400	2,400
EBT	17,600	17,600	17,600
Tax (40%)	7,040	7,040	7,040
Net Profit	10,560	10,560	10,560

Cash Amount Earned through Operation – After Tax			
Net Profit	10,560	10,560	10,560
Add Depreciation	2,400	2,400	2,400
Cash from Operations	12,960	12,960	12,960

driven, its depreciation will vary depending on the number of miles it has been driven. Let's say it was driven 15,000 miles in the first year; another 30,000 miles in the second year; and finally, 20,000 miles in the third year. Let's see how this method changes the company's performance if everything else remains the same.

This time, the vehicle's deprecation is (15,000 miles x $0.12 per mile =) $1,800 in the first year; and it changed in the second year to (30,000 miles x $0.12 per mile =) $3,600. The third year's depreciation is (20,000 miles x $0.12 per mile =) $2,400. The result of these changes in depreciation is shown on the Table PA-02.

The table above presents different amount of EBT each year. However, when no tax is considered, the cash amount earned each year remains the same at $20,000 as seen in the case of Straight Line Method. When tax is applied by 40%, the Net Income is the smallest in the 2nd year when the largest amount of Depreciation is applied. However, the amount of cash earned becomes the largest in the 2nd year when the non-cash expense is added back. This happened because the amount of tax is the smallest in the 2nd year, too. This brings up the additional benefit of depreciation for the business. By presenting a non-cash expense, the business saves on its tax expense. In summary, investing into business has multiple benefits: one is increasing its operational capacity, and the other is it ends up with saving in its tax expense which is considered real saving. The ultimate impact is the amount of the non-cash expense remains in the cash account of the company.

The next is Accelerated Method. As mentioned earlier, this method depreciates assets in their earlier years. By doing so, it presents larger amount of depreciation in the beginning years; and then the amount grows smaller over time. Since the impact of this method is the same as that which was observed in the Usage Method when the depreciation amount was the largest, the next sections will only introduce the calculation methods and the tables of projected results.

3. *Accelerated Methods*

There are two different methods in this approach. One is Double Declining method; and the other is Sum-of-the years Digits method. Each method is explained separately below.

Table PA-02 The Impact of Depreciation on Tax, Profit, and Cash – Usage Method

With no tax	Year 1	Year 2	Year 3
Revenues/Sales	1,00,000	1,20,000	1,40,000
Cost of Sales	30,000	40,000	50,000
Gross Profit	70,000	80,000	90,000
Expenses...	50,000	60,000	70,000
Depreciation	1,800	3,600	2,400
EBT	18,200	16,400	17,600

Cash Amount Earned through Operation – No Tax			
EBT (Taxable Income)	18,200	16,400	17,600
Add Depreciation exp.	1,800	3,600	2,400
Cash from Operations	20,000	20,000	20,000

With tax	Year 1	Year 2	Year 3
Revenues/Sales	1,00,000	1,20,000	1,40,000
Cost of Sales	30,000	40,000	50,000
Gross Profit	70,000	80,000	90,000
Expenses...	50,000	60,000	70,000
Depreciation	2,400	2,400	2,400
EBT	17,600	17,600	17,600
Tax (40%)	7,040	7,040	7,040
Net Profit	10,560	10,560	10,560

Cash Amount Earned through Operation – After Tax			
Net Profit	10,560	10,560	10,560
Add Depreciation	2,400	2,400	2,400
Cash from Operations	12,960	12,960	12,960

3.1. (Double) Declining Method

This method, different from all other techniques, uses the entire purchase amount of the asset in the first year; and the remaining balance (net book value) is used in the following years. In our example, the vehicle was purchased at $13,000. From this amount ($13,000), annual depreciation expense is calculated. Since the vehicle was estimated to be in use for 5 years, the vehicle is supposed to lose 1/5 of its net book value (20%) each year. As the name says, however, we are "Doubling" the loss of value – from 20% per year to 40% per year (this is what "Double Declining" means).

In the first year, the vehicle will lose 40% of its original purchase amount (0.4 x $13,000 = $5,200). With this value lost, the vehicle will be work only ($13,000 - $5,200 =) $7,800 in the second year. During the second year, the vehicle will lose another 40% of its remaining value ($7,800). The depreciation expense will be (0.4 x $7,800 =) $3,120. The vehicle's Net Book Value is only $4,680. During the third year, it will lose another 40% of its $4,680. Its annual depreciation will be $1,872, leaving the Net Book Value at $2,808. This process continues until the asset reaches its originally estimated Salvage Value at $1,000. The last year's deprecation is usually the difference between the net book value and its predetermined salvage value. Refer to the table below; and observe the impact of this technique on the tax, net profit, and cash as presented in the previous two cases.

As seen in the Usage Method scenario, the cash amount earned when no tax is considered remains at $20,000. However, when tax expense is assessed, the first year's cash amount earned is the largest although the net profit amount was the smallest in the first year.

3.2. Sum-of-Years' Digits

The other technique of Accelerated methods is the Sum-of-Years' Digits method. The only difference of this technique is the way to calculate the portion of the Asset item's annual depreciation. As its name implies the Sum (total) of the digit of each year (within which the item depreciates = Useful Life) is used to calculate.

Table PA-03 The Impact of Depreciation on Tax, Profit, and Cash – Double Declining

With no tax	Year 1	Year 2	Year 3
Revenues/Sales	1,00,000	1,20,000	1,40,000
Cost of Sales	30,000	40,000	50,000
Gross Profit	70,000	80,000	90,000
Expenses...	50,000	60,000	70,000
Depreciation	4,000	3,200	2,400
EBT	16,000	16,800	17,600

Cash Amount Earned through Operation - No Tax			
EBT (Taxable Income)	16,000	16,800	17,600
Add Depreciation exp.	4,000	3,200	2,400
Cash from Operations	20,000	20,000	20,000

With tax	Year 1	Year 2	Year 3
Revenues/Sales	1,00,000	1,20,000	1,40,000
Cost of Sales	30,000	40,000	50,000
Gross Profit	70,000	80,000	90,000
Expenses...	50,000	60,000	70,000
Depreciation	4,000	3,200	2,400
EBT	16,000	16,800	17,600
Tax (40%)	6,400	6,720	7,040
Net Profit	9,600	10,080	10,560

Cash Amount Earned through Operation – After Tax			
Net Profit	9,600	10,080	10,560
Add Depreciation	4,000	3,200	2,400
Cash from Operations	13,600	13,280	12,960

The vehicle used in this example was purchased at $13,000; and is scheduled to be depreciated for 5 years (Useful Life) to $1,000 (Residual Value or Salvage Value). Thus, the digit of each year is 1, 2, 3, 4, and 5. The sum of these five digits is (1 + 2 + 3 + 4 + 5 =) 15. This total (sum) is used as the "denominator" in the fraction to calculate each year's Depreciation Expense.

To report the largest amount of depreciation in the first year, the last digit of the years (which is "5") is used as the numerator. Thus, "5/15" is applied to the entire amount of depreciations ($12,000, which is $13,000 purchase cost – $1,000 Salvage Value) to determine the first year's depreciation ($12,000 x 5/15 = $4,000). In the second year, "4/15 x $12,000 = ($3,200)" is applied; and "3/15 x $12,000 = $2,400)" for the third year; and so forth. This way, the total amount of 5 years' depreciation ($12,000) is allocated to each year from the largest amount to the smallest gradually.

The table above shows the amounts of annual depreciation expenses. Gradually decreasing amounts of annual depreciation expenses result in growing amounts of EBT. However, if we add each year's depreciation to the EBT of the year, the cash amount obtained through operations remain the same as those that used other methods introduced earlier. So, the conclusion can be made here that regardless of the methods used, the amount of cash obtained from operations remains the same if tax is not considered. However, when tax is considered, the larger the depreciation is reported, the more tax expense is saved; and more cash amount remains in the company for its use, and for the investors.

MACRS (sounds like "makers") – Modified Accelerated Cost Recovery System

Most companies depreciate their long-term assets using one of the methods introduced in this chapter. However, to make the depreciation methods more consistent, a governmental guideline called "MACRS" has been designed. Under this guideline, assets are grouped into a few categories; and each category of assets can be depreciated within the given terms. For example, Panera Bread Co. depreciates its long-term assets as follows:

Table PA-04 The Impact of Depreciation on Tax, Profit, and Cash – Sum of the Years' Digit

With no tax	**Year 1**	**Year 2**	**Year 3**
Revenues/Sales	1,00,000	1,20,000	1,40,000
Cost of Sales	30,000	40,000	50,000
Gross Profit	70,000	80,000	90,000
Expenses...	50,000	60,000	70,000
Depreciation	4,000	3,200	2,400
EBT	16,000	16,800	17,600

Cash Amount Earned through Operation – No Tax			
EBT (Taxable Income)	16,000	16,800	17,600
Add Depreciation exp.	4,000	3,200	2,400
Cash from Operations	20,000	20,000	20,000

With tax	**Year 1**	**Year 2**	**Year 3**
Revenues/Sales	1,00,000	1,20,000	1,40,000
Cost of Sales	30,000	40,000	50,000
Gross Profit	70,000	80,000	90,000
Expenses...	50,000	60,000	70,000
Depreciation	4,000	3,200	2,400
EBT	16,000	16,800	17,600
Tax (40%)	6,400	6,720	7,040
Net Profit	9,600	10,080	10,560

Cash Amount Earned through Operation – After Tax			
Net Profit	9,600	10,080	10,560
Add Depreciation	4,000	3,200	2,400
Cash from Operations	13,600	13,280	12,960

- Leasehold improvements 15 – 20 years
- Machinery and equipment 3 – 15 years
- Furniture and fixture 2 – 7 years
- Computer hardware and software 3 – 5 years

Finally, depreciation expenses are not used to assess the value of assets. They are used to recover the Capital Investments spent to obtain the assets. For this reason, they are considered Cost Recovery System. Regarding the depreciation methods, the Consistency Principle must be applied. However, it does not mean that a restaurant (or any other businesses) must use the same method for preparing the tax report that it uses for preparing financial statements for investors. A business may use different methods for different purposes, as long as they use the same method consistently for the same purpose. Even in a company that has multi-unit operations, for example, the straight-line method be used for one unit, and an accelerated method of depreciation may be used to depreciate another unit.

INDEX

S

T

U

V

For Product Safety Concerns and Information please contact our EU representative GPSR@taylorandfrancis.com Taylor & Francis Verlag GmbH, Kaufingerstraße 24, 80331 München, Germany

Batch number: 10731081

Printed by Printforce, the Netherlands